AN INTRODUCTION TO ASTRONOMY: THE SOLAR SYSTEM

JERRY WAXMAN
SANTA ROSA JUNIOR COLLEGE

KENDALL/HUNT PUBLISHING COMPANY
4050 Westmark Drive   Dubuque, Iowa 52002

This edition has been printed directly from camera-ready copy.

Copyright © 1992 by Gerald D. Waxman

ISBN 0-8403-7753-3

All rights reserved. No part of this publication may be reproduced, stored in a retrieval system, or transmitted, in any form or by any means, electronic, mechanical, photocopying, recording, or otherwise, without the prior written permission of the copyright owner.

Printed in the United States of America

10 9 8 7 6 5 4 3 2

Table of Contents

1. Introduction: The Realm of Astronomy, 1
  The Solar System, 2
   The Sun, 3
   The Major Planets, 3
   The Minor Planets, 4
   The Satellites, 4
   The Comets, 4
   The Structure of the Solar System, 4
   The Scale of the Solar System, 4
  The Stellar System, 5

2. The History of Astronomy, 7
  The Greek Astronomers, 9
  The Renaissance Astronomers, 10
   Nicolaus Copernicus, 10
   Tycho Brahe, 11
   Johannes Kepler, 11
   Galileo Galilae, 12
   Isaac Newton, 13
   The Laws of Motion, 14
   The Law of Gravity, 14
   Orbits, 15

3. The Major Planets, 17
  Mercury, 17
   The Orbit of Mercury, 17
   The Rotation of Mercury, 17
   Observations of Mercury, 20
   Mercury and Relativity, 22
  Venus, 24
   The Rotation of Venus, 24
   The Atmosphere of Venus, 25
   The Temperature at the Venusian Surface, 25
   The Pressure at the Venusian Surface, 25
   Space Probes to Venus, 26
   The Surface of Venus, 26
  Mars, 32
   Appearance of Mars, 34
   Properties of Mars, 34
   Exploring Mars, 35
   The Surface of Mars, 36
   The Volcanoes of Mars, 36
   The Canyon of Mars, 41
   Water on Mars, 42
   The Polar Caps, 44
   The Viking Experiments, 44
   The Martian Satellites, 45

Jupiter, 47
- General Properties of Jupiter and its Orbit, 47
- Space Probes to Jupiter, 48
- The Great Red Spot, 49
- The Atmosphere of Jupiter, 49
- The "Body" of Jupiter, 51
- The Satellites of Jupiter, 52
  - Io, 52
  - Europa, 52
  - Ganymede, 55
  - Callisto, 55
- The Ring of Jupiter, 58

Saturn, 59
- The Rotation and Revolution of Saturn, 59
- The Mass and Density of Saturn, 59
- Space Probes to Saturn, 60
- The Rings of Saturn, 63
- The Satellites of Saturn, 63
  - Mimas, 64
  - Enceladus, 65
  - Tethys, Rhea, Dione, 65
  - Titan, 65
  - Hyperion, 65
  - Iapetus, 66
  - Phoebe, 66

Uranus, 67
- The Discovery of Uranus, 67
- The "Body" of Uranus, 67
- The Rings of Uranus, 69
- The Satellites of Uranus, 70
- Miranda, 70

Neptune, 72
- The Discovery of Neptune, 72
- The Orbit of Neptune, 72
- The "Body" of Neptune, 72
- The Rings of Neptune, 74
- The Satellites of Neptune, 74
- Triton, 74

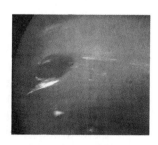

Pluto, 76
- The Orbit of Pluto, 76
- The Discovery of Pluto, 76
- The Discovery of Charon, 77
- The Mass of Pluto, 78
- The Surface of Pluto, 78
- The Atmosphere of Pluto, 80
- The "Body" of Pluto, 80

4. The Minor Members of the Solar System, 81
The Minor Planets, 81
- The Largest Minor Planets, 83
- Minor Planets Outside the Asteroid Belt, 83
- The Origin of Asteroids, 84

The Comets, 84
 The Orbits of Comets, 85
 The Structure of Comets, 85
 The Origin of Comets, 89
 The Naming of Comets, 89
Meteors and Meteoroids, 90
 Sporadic Meteors, 90
 Meteor Showers, 90
Meteorites, 93
Meteorite Types, 94

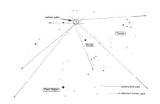

5. The Moon, 95

The Orbit of the Moon, 95
The Month, 96
The Orbit Plane of the Moon, 96
The Phases of the Moon, 97
The Distance to the Moon, 101
The Physical Properties of the Moon, 101
The Rotation of the Moon, 102
The Diameter of the Moon, 102
The Mass of the Moon, 103
The Lunar Surface, 104
The Lunar Surface Gravity, 112
The Lunar Atmosphere, 112
Water on the Lunar Surface, 112

The Lunar Surface Temperature, 113
The Lunar Space Program, 113
 The Ranger Series, 113
 The Surveyor Series, 114
 The Lunar Orbiter Series, 116
 The Mercury Series, 116
 The Gemini Series, 116
 The Apollo Series, 116
 The Results from Apollo, 118
The Lunar Interior, 119
The Far Side of the Moon, 120

The Origin of the Moon, 120
 The Fission Theory, 120
 The Capture Theory, 120
 The Accretion Theory, 121
 The Impact Theory, 121
Eclipses, 122
 Solar Eclipses, 123
 Lunar Eclipses, 126
 Frequency of Eclipses, 128
 Eclipse Cycles, 128
 Visibility of Eclipses, 129
Tides, 129

6. The Sun, 133
    How We Study the Sun, 133
    The Electromagnetic Spectrum, 134
    Spectroscopy, 135
    The Solar Photosphere, 137
        Photospheric Granulation, 138
        Sunspots, 138
        The Sunspot Cycle, 141
    The Solar Chromosphere, 144
    The Solar Corona and the Solar Wind, 144
    The Solar Interior, 148
        Nuclear Fusion, 148
        The Future of the Sun, 149

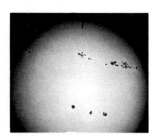

7. The Origin of the Solar System, 151
    Before the Beginning, 152
    The Early Solar System, 153
    The Formation of the Planets, 154
    The Formation of Satellites, 155
    The Formation of Comets and Asteroids, 155

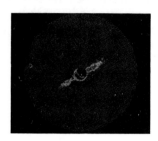

Appendices
    Appendix I: Planetary Orbital Data, 157
    Appendix II: Planetary Physical Data, 159
    Appendix III: Satellite Data, 161
    Appendix IV: Learning Activities, 163
    Learning Activity 1:   The Geocentric Hypothesis and Planetary Motion, 163
    Learning Activity 2:   Retrograde Motion in Heliocentric Orbits, 165
    Learning Activity 3:   Kepler's 1st Law: The Ellipse, 167
    Learning Activity 4:   Kepler's 2nd Law: Law of Areas, 169
    Learning Activity 5:   Kepler's 3rd Law: Harmonic Law, 171
    Learning Activity 6:   Newton's Law of Gravity; Surface Gravity, 173
    Learning Activity 7:   The Orbit of Mercury, 175
    Learning Activity 8:   The Semi-Major Axis of Venus and the Determination of the Au, 179
    Learning Activity 9:   The Mass of Mars, 181
    Learning Activity 10:   Mass of the Pluto/Charon System, 183
    Learning Activity 11:   Radiant Point of Meteor Shower, 185
    Learning Activity 12:   Time from Lunar Phases, 187
    Learning Activity 13:   The Sunspot Cycle, 189

    Appendix V: Star Charts
    Equatorial Charts (fall/winter), 191
    Equatorial Chart (spring/summer), 193
    North Polar Chart, 195

Credits, 197

Index, 199

Preface

In recent years, astronomy has become so broad that it is now impossible, in a single semester, to impart to a beginning student the full sweep of the field. This has led many colleges and universities to adopt the practice of teaching astronomy in two parts. Generally, the division is placed at the boundary between our solar system and interstellar space, so that one course would concern the solar system while the other would cover everything which lies beyond. Only in this way is enough class time available to do justice to this oldest of sciences.

A review of available texts, however, reveals that almost all cover astronomy from A to Z and, moreover, in such an encyclopedic fashion that most students cannot hope to read, let alone master, even half of it in a single semester. It has been my experience that students, when presented with such a book, become overwhelmed before they begin.

This two volume set is an attempt to remedy this situation. The first volume, "The Solar System", covers just that and only that. It is suitable for a freshman level, one semester, solar system course. Moreover, the material content has been limited to that which one might reasonably expect students to master; it is a text, not an encyclopedia. Similarly, the second volume, "Stars, Galaxies and the Universe", covers the basic astronomic properties of the universe beyond the solar system. As with Volume I, it is limited in scope and manageable in content. I believe that the use of these texts not only saves the students money (an item which should not be overlooked these days), but gives the student a better chance at success since he or she is not being given more material than can possibly be covered in class or read at home.

Additionally, since one remembers best what one experiences first hand, Appendix IV contains about a dozen "learning activities"; short homework or classroom exercises which serve to illustrate particular important principles discussed in the text. These exercises are complete and require no special equipment. I have found that students, having completed exercises such as these, have a much firmer grasp of the course material.

Jerry Waxman

Acknowledgments

The author wishes to acknowledge and thank the following people whose assistance was invaluable in the writing of this book:

Keith Waxman and Ron Smith; friends and colleagues whose editorial comments prevented me from appearing totally foolish (I hope!).

Marita Gardner, whose hours of secretarial work literally created the book that you are holding.

Pam Zimmerman, my partner and friend whose understanding and support enabled me to succeed in a project which often had its difficult moments.

Thank you all.

CHAPTER 1

INTRODUCTION: THE REALM OF ASTRONOMY

Before attempting to recount the details of any field of endeavor, it is important to first summarize the generalities of that subject so that the student may conveniently incorporate these specifics into a pre-existing framework. Particularly, this should be done for "first-time" astronomy students as the material to be presented is beyond the realm of experience and, hence, often difficult to digest. This chapter, then, is an attempt to give an abbreviated picture of the modern scientific view of our universe. In it, we will try to describe the "cast of characters" which populate the universe and give the student a feeling for his/her position or location within the cosmos. Once the general nature of the universe is understood, the specifics to be encountered later will be more readily digested.

Thousands of years ago, our ancestors viewed the heavens with awe and mystery. To these people, the sky was filled with beings who possessed great power; gods and goddesses who looked down upon the Earth and who controlled its destiny with casual disregard. The manifestations of these powerful entities, the **Sun**, **Moon**, **planets** and **stars**, moved through the heavens in complex, undecipherable patterns which bewildered our forebearers and supported the contention that the "universe" could never be predictable since each and every happening was the direct result of the capricious act of an independent deity.

All of this changed, however, 2500 years ago on the isles and backwaters of the Aegean Sea, where ancient Greek scientist/philosophers had an idea that was to change the world. They decided that the world was not ruled by the random whims of fickle gods, but rather by immutable **laws of nature** which constrained events in the physical universe. By suitable study, it was believed, one might come to know these laws and hence predict future happenings.

Over the next 600 years such men as **Pythagoras**, **Aristarchus**, **Eratosthenes**, **Plato**, **Aristotle**, and **Ptolemy** (all to be discussed in the next chapter) sought, through contemplation and observation, to divine the rules by which the universe conducts its activities. These initial attempts at "scientifically" describing the universe set us on the road to our modern technological society. This road, however, was not unbroken.

After the fall of the Greco/Roman civilization, 1500 years of scientific "darkness" had to pass before the re-awakening of thought brought about by **Copernicus**, **Galileo**, **Kepler**, **Brahe**, **Newton** and many others. These 16th and 17th century scientists "picked up" where the Greeks had left off. Only now, armed with a new scientific method based on experiment and observation, the discoveries came at a furious rate. First the Sun, which heretofore had been thought to orbit the Earth, was placed in its correct position at the center of the Solar System. Then the true motions of the planets and shapes of their orbits were discovered. The invention of the telescope led to our understanding of gas clouds and star clusters in the depths of space beyond the Solar System.

The invention of "**spectroscopy**" and the realization of the existence of wavelengths of light beyond the visible opened a chapter in astronomy which led to an understanding of the births and deaths of stars, the discovery of other galaxies and the observation of the first moments in the life of the universe.

In the last 2500 years, astronomy has gone from a loosely organized set of half-truths and misconceptions based on mysticism and superstition to an objective science based on precise observations and verifiable theories. This science has led to a remarkable understanding of our universe; an understanding which we will attempt to communicate in the following chapters of this book.

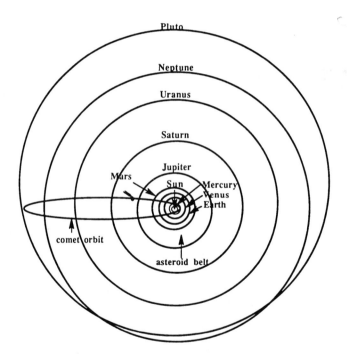

Figure 1.1: The Solar System

THE SOLAR SYSTEM

The human species occupies a small planet orbiting a mundane star known as the **Sun** or **Sol**. This star controls the destinies of a host of smaller bodies known as planets, satellites and comets. Taken as a group, the Sun and its family is known as the **Solar System** or the system of **Sol**. The general layout of the Solar System is shown in Figure 1.1 (the layout of the figure is <u>not</u> to scale).

The Sun: The Sun, which lies at the center of the Solar System, is a rather typical star (it is larger and brighter than average, but not too much so). It is just like the thousands of stars which we see in the night sky and looks larger, brighter and feels hotter only because we are nearer to it. The other stars, were you as close to them as you are to the Sun, would appear as majestic.

The **stars** (Sun included) may be superficially described as "big balls of hot gas" or, if you like illiterations, as "great globes of glowing gas". The Sun itself is a sphere of hydrogen gas, with a surface temperature of approximately 11,000° F and a diameter of about 860,000 miles. Over 100 "Earths", placed side to side, would be required to cross the diameter of the Sun (Figure 1.2).

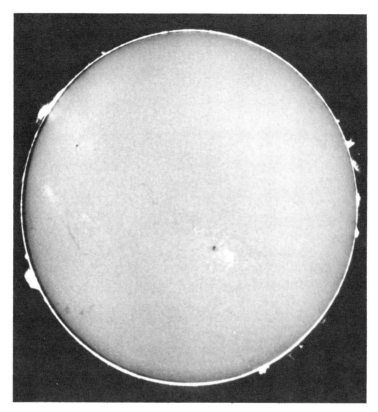

Figure 1.2: The Sun

As far as we know, the Sun is the only star in the Solar System. All of the other stars are at immense distances; hence their dim appearances. The Sun will be discussed in detail in Chapter 6.

As described by Isaac Newton in the 17th century (Chapter 2), all material objects in the universe exert a "force of **gravity**" on all other material objects. This force is greater for objects containing more mass (matter) and for objects which are closer together. The Sun, having a large amount of mass (it is the equivalent of 333,000 "Earths"), is the dominant source of gravity in this part of the universe and thus is able to "hold" in its grip a myriad of smaller bodies, chief of which are the planets.

The Major Planets: **Planets** may be <u>generally</u> defined as relatively low mass (less than 0.08 times the mass of the Sun), relatively cool, bodies orbiting stars. Our star, the Sun, possesses nine major planets whose orbits are shown in Figure 1.1. The planet on which we live, the **Earth**, is the third in distance from the Sun. The orbits of the planets are elliptical, but only so slightly as to appear (except in two cases) as circles. The diameters of the major planets range from 1400 miles (Pluto) to 88,000 miles (Jupiter). The Earth is about 8,000 miles across. The major planets will be discussed in Chapter 3.

The Minor Planets: In the space between Mars and Jupiter are found thousands of small orbiting bodies called **minor planets, asteroids** or **planetoids**. The largest of these, Ceres, is about 600 miles in diameter. These bodies will be discussed in Chapter 4.

The Satellites: Orbiting about most planets, trapped by their gravitational pulls, are smaller bodies called **satellites** (sometimes **moons**). All together, there are at least 60 satellites orbiting the planets of the Sun. Some of these satellites are as large as the smaller major planets but most are more like the minor planets in size. The Moon will be discussed in Chapter 5.

The Comets: Orbiting the Sun in long elliptical paths are the **comets** (Figure 1.1). These objects are balls of ice, rock and dust from 1/4 mile to 20 miles in diameter. When the comet is far from the Sun it is frozen and invisible to Earth scientists. As it approaches the Sun, however, the solar heat vaporizes the ices and pressure from the Sun pushes the vapors into space, forming the **tail** of the comet. As the comet again moves into deep space, it freezes and becomes invisible until its next solar passage. Comets will be discussed in Chapter 4.

The Structure of the Solar System: As one can see from Figure 1.1, the Solar System is a very "flat place". The orbits of the planets are co-planer. The greatest deviation from this rule is the planet Pluto, whose orbital inclination to the plane of the Solar System is 17°. Also, the planets all move about the Sun in the same direction (counter clockwise when viewed from the north). The comets, however, have more random orbits; some lie close to the plane of the planetary orbits but most show high inclinations and their directions of revolution are also random. The co-planer nature of the planetary orbits bespeaks the origin of the Solar System and the chaotic nature of the comet orbits tells us of their origin (The origin of the Solar System will be discussed in Chapter 7).

The Scale of the Solar System: To get an idea of the size of our Solar System, we must first understand the units which astronomers use to measure cosmic distances. "Normal" distance measuring units used on Earth (e.g. miles, kilometers) are not convenient for astronomical distances because these units are too small; the number of miles between the planets or the stars is so large that just writing them is cumbersome and inconvenient.

The Astronomical Unit: As it is permissible to define any length unit which is convenient, astronomers have decided to choose one in which the distance from the Earth to the Sun is called 1. Formally, **the astronomical unit (au)** is defined as "the mean Earth-Sun distance". The length of the "astronomical unit" is approximately 93 million miles.

Units of Light Travel Time: Another method of describing large distances is to state the time required for light to travel these distances. Since the speed of light, a constant*, is approximately 186,000 miles/second, a distance of 186,000 miles would be equivalent to 1 **light second**. One **light minute**, the distance light travels in a minute, is 60 times as large as a light-second or about 11 million miles. A **light hour** is equal to 60 light minutes and a **light day** is 24 light hours. Finally, a **light year** is 365 light days (A light year is a distance of about 6,000,000,000,000 miles!). All of these "light" units are units of distance rather than time.

Figure 1.3 is a chart which shows the scale of the Solar System in units of miles, au's, light-travel units and, in the last column, the time it would take to drive the distance at 100 miles per hour. This last column is included to give you a "feeling" for the large distances between the planets of the Solar System.

Another way to appreciate the distances between bodies in the Solar System is to construct a "scale model". If we were to scale down the Sun so that it is one foot in diameter (its true diameter is approximately one million miles), the major planets in the Solar System would have the diameters and distances from our "model" Sun shown in Figure 1.4.

*The speed of light is constant in a vacuum.

Distance from	Miles	Au's	Light Travel	@ 100 mph
Earth to Moon	240,000	1/400	1.3 light seconds	100 days
Earth to Venus (closest planet)	25,000,000	0.27	2.2 light minutes	29 years
Earth to Sun	93,000,000	1.00	8.3 light minutes	106 years
Sun to Pluto (radius of Solar System)	4,000,000,000	40	5 light hours	4240 years

Figure 1.3: Distances in the Solar System

Object	Diameter (inches)	Distance from Sun
Sun	12	-
Mercury	0.04	35' (10 yards)
Venus	0.11	67' (20 yards)
Earth	0.11	100' (30 yards)
Mars	0.06	150' (50 yards)
Jupiter	1.2	500' (0.1 mi)
Saturn	1.0	1000' (0.2 mi)
Uranus	0.4	2000' (0.4 mi)
Neptune	0.4	3000' (0.6 mi)
Pluto	0.02	4000' (0.8 mi)

Figure 1.4: Scale Model of the Solar System

THE STELLAR SYSTEM

Beyond the Solar System is a vast sea of emptiness. If one were to travel at the speed of light through this space, after 4.3 years he/she would arrive at the next nearest star system; the Alpha Centauri triple star system. This system is approximately 25,000,000,000,000 (25 trillion) miles from the Solar System. At the speed of an Apollo moonship, it would take 100,000 years to reach this nearest star system. If it were placed in the scale model of the Solar System mentioned previously, it would be 5000 miles away from our basketball sized Sun!

On average, the star systems in our part of space are five to seven light years apart (30 to 40 trillion miles) and the spaces between them, though sometimes occupied by large tenuous gas clouds called "nebulae", are essentially devoid of all matter.

Although only about 6,000 distant star systems can be seen with the naked eye from the surface of the Earth, many more actually exist. In the early part of this century it was ascertained that about 100,000,000,000 (100 billion) star systems are found within 80,000 light years of our location. These star systems are constrained and contained by the combined effect of all their individual gravities. Taken together, these star systems form the **Milky Way galaxy**. The visible portion of this galaxy is approximately 100,000 light years across, and its plane (edge) can be seen glowing dimly on a dark summer night (Figure 1.5).

In the 1920's, astronomers at Mt. Wilson observatory discovered that our Milky Way galaxy was only one out of billions of such systems which populate **intergalactic space**. These galaxies are hundreds of thousand to millions of light years apart and have been seen out to distances in excess of 10 billion light years. Taken together, all of these galaxies make our "visible" **universe** (Figure 1.6).

In the last few decades, astronomers have embarked upon a study of the origin of the universe. Studying the history of the systems of galaxies is aided by the fact that looking into the distance is equivalent to looking into the past. This is true because of the finite velocity of light (e.g. if, tonight, we observe a galaxy that is 1 million light years away, we are seeing it as it was 1 million years ago since that's how long it takes its light to reach us). This effect has allowed us to look back to the apparent time of the creation of the universe. These observations have helped us model and refine our theories.

Presently, astronomers are making observations which will allow them to predict the destiny or future characteristics of our universe. These studies are perhaps the most dramatic of all the scientific pursuits undertaken by humans and speaks of the great intellectual powers possessed by this physically frail life form populating the mote of dust called Earth.

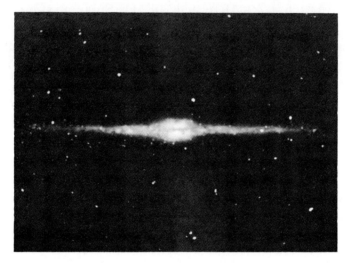

Figure 1.5: The Milky Way Galaxy

Figure 1.6: A cluster of galaxies in intergalactic space

CHAPTER 2

THE HISTORY OF ASTRONOMY

From the earliest times, humans have held a fascination for the sky. Over 5000 years ago, ancient navigators learned that they could find their way back home by sailing in the direction of particular stars or patterns of stars. The first **constellations** (figures in the sky formed by connecting bright stars with straight lines) may have been an attempt by these navigators to remember particular important navigational stars as it is easier to recollect line patterns than disconnected "dots".

It was also noticed that five "stars" had the ability to move among the rest which remained "fixed" relative to one another. These five "wanderers" were thought to have great powers and were elevated to the status of gods by the Egyptians and later by the Greeks and Romans. [The Greek word for wanderer is πλανητησ (planets)]. The constellations through which the planets moved (there are 13 of them) were also considered to have great significance and were collectively known as the ζωδιακ (zodiac or zone of animals). (Figure 2.1).

The notion developed that since the planets were gods and therefore controlled the destiny of the Earth and its inhabitants, one might be able to predict the future by noting the position of a planet relative to the constellations of the zodiac. For example, if a certain event took place the last time Mars was in Taurus (famine, flood, death of a ruler, etc.) maybe the same event would occur the next time Mars entered that constellation. The trick was first to be able to predict when certain planets would enter and leave certain constellations and second, to know what these events predicted.

The prediction of planetary movements was complicated by the fact that the normal slow eastward (prograde) motion of the planets (against the background of fixed stars) was interrupted at seemingly random times by reversals of direction (retrograde motion). Since the times of

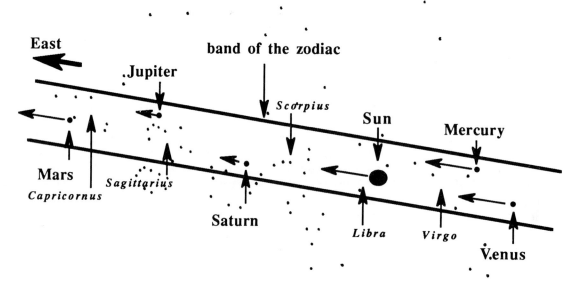

Figure 2.1: The Zodiac

retrograde motion could not be predicted, astrologers were unable to make long term prognostications.

Around 600 BC a new thought developed in the region of the world known as Greece. The idea was that the world was <u>not</u> controlled by the capricious whims of gods and goddesses but rather by immutable **laws of nature**. These laws governed the way that the universe worked and man, using suitable methods, could come to learn these laws and hence predict happenings in the physical world.

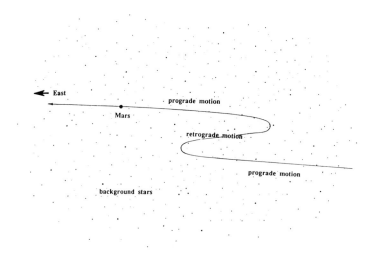

Figure 2.2: Retrograde Motion of Mars

Over the next 800 years, a series of Greek philosopher/scientists, using a mixture of observation and rational thought brought astronomy to a new height. While their concepts are often not the same ones we have today, the Greek rejection of mysticism, superstition and religion as roads to the truth, paved the way to our modern technological society. What follows are some of the most important Greek scientists and their contributions and ideas.

Pythagoras (580-500 BC): Possibly the first person to recognize that the Earth was a sphere and who coined the phrase χοσμοσ (cosmos) for the heavens above.

Parmenides (504-450 BC): Recognized that the Moon did not shine with light of its own but rather by reflected light.

Anaxagoras of Clazomenae (500-428 BC): Stated that the Sun was a glowing mass and that the Moon was made of rock and soil. He explained the phases of the Moon and eclipses.

Democritus of Abdera (~400 BC): Explained that the Milky Way was an assemblage of many faint and distant stars.

Plato (~400 BC): May have been the first to suggest that the Earth rotated on its axis. He believed that the celestial bodies were divine and, therefore, must move in perfect circles with uniform speed. He believed that anything less would not be suitable for gods. According to Plato, "reality" consisted of ideas and the "sensory" world was an illusion. "Truth" could only be found from rational thought, not from observation.

Eudoxos of Knidos (408 to 355 BC): Developed a model of the crystalline spheres which supposedly carried the planets in their courses about the Earth, which was considered to be at the center of the Universe.

Aristotle (384 - 322 BC): A student of Plato, he mirrored the ideas of his master except that he believed that "reality" was to be found in visible phenomena and observation. He also believed in experimentation. He believed, as did Plato, that the Universe was perfect and, therefore, had to be spherical. The celestial bodies were spherical in shape and moved on crystalline spheres. The universe beyond the sphere of the Moon was changeless; all observed changes took place on or near the Earth.

Heraklides of Pontus (388 - 315 BC): Proposed that Mercury and Venus moved in circles around the Sun, while the Sun moved about the Earth. He introduced the idea of **epicycles** and **deferents** to explain retrograde motion (Figure 2.3).

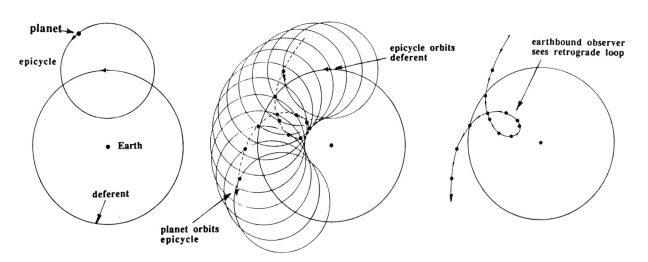

Figure 2.3: Epicycles and Deferents

Aristarchus of Samos (310 - 230 BC): Placed the Sun at the center of the universe with the Earth and other planets in orbit about it. Had the Earth rotating on its axis and realized that the Earth's equator was inclined to the plane of its orbit. Estimated the Sun to be seven times the diameter of the Earth.

Eratosthenes of Syene (270 - 190 BC): Measured the circumference of the Earth to be about 25,000 miles (within 2 percent of the correct value). He did this by noticing that on June 22 (the longest day of the year) the Sun, while overhead at the town of Syene, was 7° from the zenith of Alexandria (Figure 2.4).

$$\frac{\text{Circumference}}{500 \text{ miles}} = \frac{360°}{7°}$$

$$\text{Circumference} = 500 \text{ miles} \times \frac{360°}{7°}$$

$$\text{Circumference} = 25,000 \text{ miles}$$

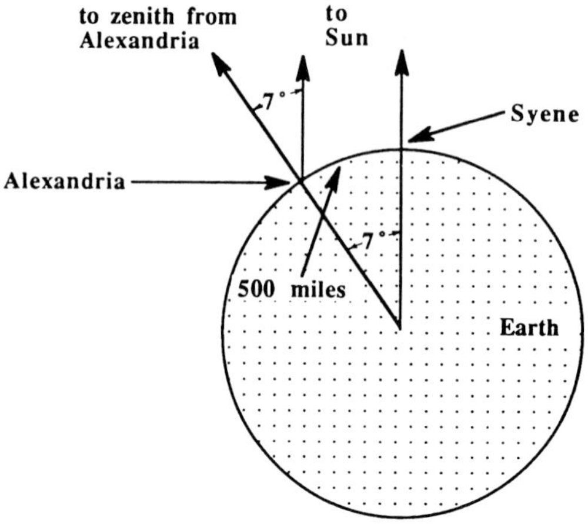

Figure 2.4: Eratosthenes' Determination of the Circumference of the Earth

Since 7° is approximately 1/50 of a circle and since Syene and Alexandria were measured to be about 500 miles apart, the circumference of the Earth must be 50 x 500 miles or 25,000 miles.

Hipparchus of Rhodes (160-125 BC): Developed the geocentric (Earth centered) hypothesis to its highest form, measured the length of the year, discovered the precession of the equinoxes, developed the scale of stellar magnitudes and expanded the work on epicycle theory.

Claudius Ptolemaeus (~150 AD): Last of the great Greek astronomers. He believed that the Earth was at the center of the universe and did not move; it did not even rotate. He developed and refined the system of epicycles and deferents into a model which agreed quite well with reality (as far as predicting planetary and retrograde motion is concerned). His system of perfect circles and uniform motions was considered "law" for fourteen hundred years, until the great "re-awakening" of the 16th and 17th centuries (Learning Activity 1, Appendix IV).

THE RENAISSANCE: By the 11th Century, the predictions of the "Ptolemaic model" were sadly in error. Although no one knew it at the time, the problem was not that the center of the system was the Earth, but rather that the orbits of the planets were taken to be circles and the motion to be uniform (constant speed). Both of these ideas were wrong.

Nicolaus Copernicus (1473 - 1543 AD): Born in Torun, Poland, he attended the University of Krakow (Poland) and the University of Padua (Italy). Earned degrees in canon law, medicine, mathematics and astronomy. Placed the Sun at the center of the Solar System. Made a "scale model" of the solar system showing the planets at their correct _relative_ distances. Explained retrograde motion as a consequence of the more rapid orbital speeds of planets closer to the Sun [i.e. as the Earth, moving faster in its orbit, passes a "superior" planet (Mars, Jupiter or Saturn), the planet appears to "back up" (Figure 2.5 and Learning Activity 2)]. Retained epicycles, "perfect" circles and uniform motion. His classic work **"On the Revolutions of the Celestial Orbs"** was published in the year of his death (Learning Activity 2, Appendix IV).

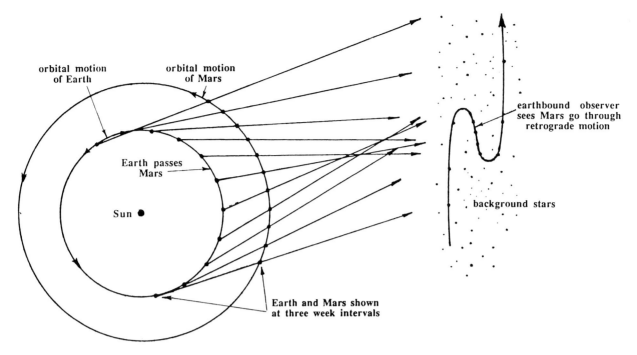

Figure 2.5: Retrograde Motion and the Heliocentric Hypothesis

The theory of Copernicus caused great controversy, particularly since it was in violation of the teachings of Aristotle, whose principles were considered infallable by the church. Taking the Earth from its "central" location was considered heresy. Add to this the fact that the Copernican system made predictions only slightly better than the Ptolemaic system (since Copernicus retained perfect circles and uniform motion - the real "culprits") and we see that acceptance of the "new world model" would not come easily.

Tycho Brahe (1546 - 1601): Originally sent to Leipzig to study law but his interest strayed to mathematics and astronomy. Observed the supernova of 1572 and demonstrated that it was at the distance of the "celestial orb", thus casting doubt upon the Greek concept of the "permanence" of the Heavens. In 1581 he was given an observatory ("**Uraniborg**") on the Danish Island of Hveen by Fredrick II, King of Denmark. Here he made the most precise observations of planetary motions up to that time. He hoped to use these positions to correct the Copernican hypothesis so that it would give more precise predictions of planetary motions. In 1599 became Royal Mathematican to Emperor Rudolph in Prague. In 1600 he was joined in Prague by Johannes Kepler, a young mathematican. He died in 1601.

Johannes Kepler (1571 - 1630): Born in Weil der Stadt, Austria, he attended University of Tübingen where he developed an interest in mathematics and astronomy. Became mathematics teacher in Graz, Austria in 1594. In 1595 he developed a theory that the planetary distances from the Sun were explained by inscribing "regular" polygons inside orbital circles (Figure 2.6). This geometrical fantasy was to plague his thoughts for the rest of his life. In 1600, he joined Tycho Brahe in Prague to help him develop a revised "world theory" which would account for planetary motion.

After Brahe's death in 1601, Kepler was appointed Imperial Mathematician by Emperor Rudolph. He observed the supernova of 1604 and stated that Aristotle had been wrong in asserting that the "Celestial Orb" was changeless. He used Brahe's observations of planetary positions (particularly Mars) to develop his 3 laws of planetary motion and to finally reconcile the Copernican hypothesis with observation.

First Law: The orbits of the planets are ellipses with the Sun at one focus of the ellipse (Figure 2.7 and Learning Activity 3, Appendix IV).

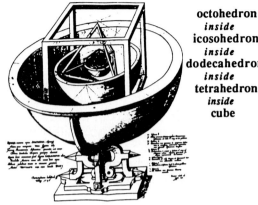

Figure 2.6: Kepler's "Nested" Polygons

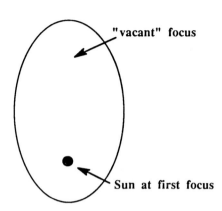

Figure 2.7: The Ellipse

Second Law: A line from the Sun to a planet sweeps out equal areas in equal times (Figure 2.8 and Learning Activity 4, Appendix IV).

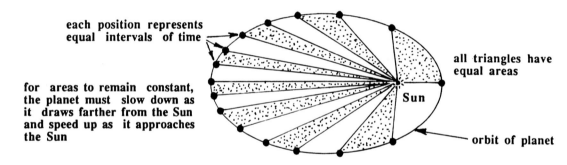

Figure 2.8: The "Law of Areas"

Third Law: The squares of the sidereal periods of the planets are proportional to the cubes of their mean distances from the Sun ($p^2 \sim a^3$). (Exercise 2.4 and Learning Activity 5, Appendix IV).

Galileo Galilae (1564 - 1642): Attended University of Pisa Medical School and stayed on as a lecturer in mathematics. In 1592 he took the position of chair of mathematics at the University of Padua. In 1610 he moved to Florence as the chief mathematician and philosopher to the Grand Duke Cosimo II of Tuscany, his former pupil. In 1609 Galileo made his first telescopic observations. He observed the Sun, the Moon, Venus, Jupiter and Saturn. In each case he made great astronomical discoveries and in each case these discoveries led him to more firmly reject the geocentric hypothesis and the teachings of Aristotle (as well as the scientific dogma of the Roman Catholic church).

Galileo's Telescopic Observations;

The Sun: Galileo found the Sun to possess "spots". He correctly deduced that the spots (called **sunspots** today) were blemishes attached to the surface of the Sun. This contradicted the notion that the Sun was "perfect". Galileo was also able to compute the rotation rate of the Sun by observing the sunspots. If the Sun rotated, it seemed to be behaving somewhat like an ordinary celestial body.

The Moon: The telescope revealed craters on the Moon. In fact, Galileo saw mountains on the Moon, and also what he thought to be seas and cities. This made the Moon far less than the perfect sphere envisioned by Plato and diminished the "unique" position of Earth.

Venus: The most important discovery made by Galileo was that the planet Venus moved through a series of phases quite similar to the lunar phases. The phases of Venus could only be explained if Venus were in orbit about the Sun rather than around the Earth. The reason for the phases of Venus is shown in Figure 2.9. The fact that Venus was in orbit about the Sun was a direct contradiction to the geocentric hypothesis. Galileo was thus convinced of the validity of the Copernican hypothesis.

Jupiter: Observations of Jupiter led Galileo to discover 4 satellites orbiting the giant planet. These satellites (**Io, Europa, Ganymede** and **Callisto**) are now known as the **Galilean satellites** of Jupiter. Galileo's observations led him to conclude that the four new bodies were orbiting Jupiter rather than the Earth. This was clearly a contradiction of the geocentric hypothesis.

Saturn: When observed through Galileo's telescope, the planet Saturn showed projections on both sides. Galileo wrote that "Saturn has 'ears'". Of course, he had discovered the rings of Saturn without knowing it. The true nature of the projections was not appreciated until about 50 years later.

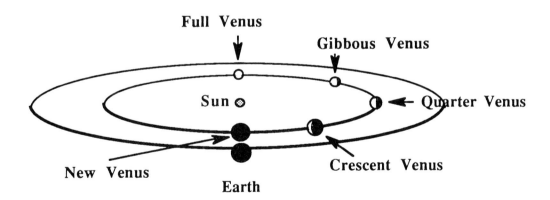

Figure 2.9: The Phases of Venus

For teaching the Copernican hypothesis, Galileo was twice brought before the inquisition and was finally, in 1633, sentenced to house arrest where he spent the last nine years of his life.

Isaac Newton (1642 - 1727): In 1642, the year of Galileo's death, Isaac Newton was born in the village of Woolsthorpe, shire of Lincoln, England. He attended Cambridge University and stayed on as a professor of mathematics. In 1665, while he was still a student, the University was closed for 18 months because of the plague sweeping London, 70 miles to the south. Newton returned home to Woolsthorpe Manor to await the reopening of his school. During this 18 month period Newton developed some of the most important principles of physics as well as making

fundamental discoveries concerning the nature of light. Among his achievements during this period were;
- development of the "laws" of motion
- conception of the "law" of gravity
- invention of calculus
- discovery of the spectral nature of light

Any one of these achievements would have marked a man for everlasting fame. Together, they form the most glorious period of scientific insight in the history of physics.

THE LAWS OF MOTION: Newton realized that the laws which govern the motions of bodies, at or near the surface of the Earth, are the same as the laws which govern the motions of celestial bodies. This point had escaped <u>all</u> who preceded him. He was able to condense the concept of motion into three simple statements.

- **First Law:** A body in uniform motion remains in uniform motion and a body at rest remains at rest, unless acted upon by a net external force.

- **Second Law:** A force acting on a mass will cause the mass to accelerate in the direction of the force (**F = ma**).

- **Third Law:** For every action (force) there is an equal and opposite reaction (force).

These statements (Laws of Motion) can be used to describe and explain all terrestrial and celestial motion. All that was needed was a statement of the forces acting on the body in motion. For celestial bodies, Newton identified the force as **gravity**.

- **The Law of Gravity:** Newton realized that the same phenomenon which caused apples to fall from trees also caused the Moon to fall around the Earth (orbit the Earth). He called this force **gravity** and defined it thusly:

"Between <u>any</u> two masses in the universe there exists a force of attraction (gravity) whose magnitude is directly proportional to the product of the masses and inversely proportional to the square of the distance between the <u>centers</u> of mass.

Stated mathematically, the law can be written as:

$$F = \frac{G\, m_1 m_2}{D^2}$$

Where **F** = force of gravity
G = constant of gravity
m_1 = mass of body 1
m_2 = mass of body 2
D = distance between <u>centers</u> of m_1 and m_2

Newton theorized that the force which causes bodies to "fall" to Earth does not stop at the tops of trees nor at the top of the atmosphere, but rather extends, albeit diminishing, to the Moon and beyond. He further theorized that the magnitude of this force decreases as the <u>square</u> of the distance from the Earth's center so that someone 2 Earth radii from the center of the Earth would feel <u>1/4</u> the gravity that would be felt at the Earth's surface (1 Earth radius from the center). In this way, all bodies feel the gravitational effects of all other bodies in the universe, although close (and more massive) bodies tend to dominate. Since the closest, most massive body is the Earth, it is this force which dominates our lives. The force of gravity between you and Earth is called your **weight** (Learning Activity 6, Appendix IV).

Orbits: Newton realized that the laws of motion and the law of gravity, acting together, predicted the phenomenon of orbits. It works like this. Suppose we drop an object (a baseball, for example) from the top of a building or from a cliff. The object, feeling the force of gravity, will be accelerated in the direction of that force (second law of motion). Near the surface of the Earth, the magnitude of that acceleration is 32 ft/sec/sec (this means that for each second of fall, the velocity will increase by 32 ft/sec). During the first second, the object will fall 16 feet (can you think of why this is true?).

As one gets farther and farther from the Earth, the force of gravity decreases. At the distance of the Moon from the Earth (60 Earth radii, center to center) the force of gravity is 1/3600 (1/60 squared) as strong as at the surface of the Earth.

Now suppose we take some baseballs to the top of a tall mountain (Figure 2.10).

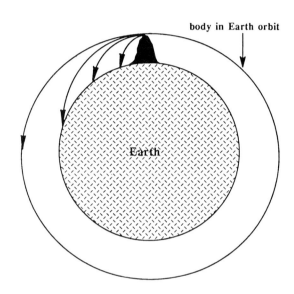

Figure 2.10: Putting a Baseball into Orbit

If we were to drop a ball vertically, it would accelerate downward at 32 ft/sec/sec, fall 16 feet in the first second and eventually hit the ground. If we were to throw it horizontally, however, it would not land at the base of the mountain because, even though it is still accelerating downward at 32 ft/sec/sec, its horizontal motion <u>persists</u> and carries it away from the base of the mountain (first law of motion). The harder we throw it horizontally, the farther it lands from the base of the mountain, even though it is always accelerating vertically at 32 ft/sec/sec.

But what if we throw it so hard horizontally that the curve of its trajectory is parallel to the curve of the Earth's surface. The ball would continue to move horizontally (first law) and continue to accelerate downward (second law) but would never get closer to the surface because the surface is curving "away" from the ball as fast as the ball is dropping! The ball is in orbit about the Earth.* To put a ball "in orbit" near the surface of the Earth, a horizontal speed of 5 miles/sec is required. Every second the ball would "fall" 16 feet but every 5 miles, the Earth's surface "curves" downward by 16 feet.

The principle stated above is the same for the Moon. At the distance of the Moon, the acceleration of gravity is 1/3600 its value at the Earth's surface. This means that, if dropped towards the Earth, the Moon would accelerate downward at a rate of 32 ft/sec/sec ÷ 3600 = 1/10"/sec/sec and, during the first second of fall, would drop 1/20".

*In this discussion, air resistance has been ignored.

But the Moon has a horizontal motion of 2/3 mi/sec [the Moon takes 27d = 2.3 million seconds, to travel $2\pi r = 2\pi (240{,}000 \text{ mi}) = 1.5$ million miles] which carries it over a place on Earth which is 1/20" farther away because of the Earth's curvature (Figure 2.11). The Moon is falling, but never can reach the Earth; it is in **free-fall**!

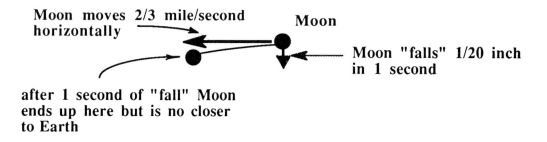

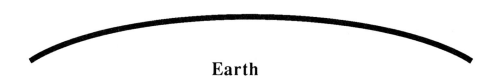

Figure 2.11: The Orbit of the Moon.

All other orbits (Earth about Sun, Sun about galactic center, artificial satellite about Earth, etc.) follow the same principle.

CHAPTER 3
THE MAJOR PLANETS

There are nine major planets orbiting the Sun. In this Chapter we will address each of these planets and discuss their gross physical and orbital properties (Appendix I).

MERCURY

Mercury is one of the brightest objects in the sky yet, because of its proximity to the Sun, most people have never seen it (its maximum angle of elongation is 28°). When it is east of the Sun, it can just barely be seen in the sunset colors for a few nights and when it is west of the Sun, it is seen in the sunrise colors. Early observers called it Hermes when it was seen as an evening object and Apollo when it was visible in the morning sky (Figure 3.1).

The Orbit of Mercury; Mercury is the planet closest to the Sun; its average distance is 35 million miles. Its orbital path, however, is quite elliptical. Its close approach to the Sun (perihelion point) is 28 million miles and its most distant point from the Sun (aphelion point) is 42 million miles. Only Pluto has a more eccentric orbit (Figure 3.2 and Learning Activity 7, Appendix IV). Mercury has no known satellites.

The plane of Mercury's orbit is tilted by 7° to the plane of the ecliptic. Only Pluto has a more inclined orbit. The sidereal period of revolution of Mercury is 88d (1/4 year). No other planet has so short a revolution period. Mercury is the only planet whose orbital plane lies exactly on the Sun's equatorial plane.

The Rotation of Mercury; Observations of the surface of Mercury are very difficult to make from the surface of the Earth. This is because of Mercury's proximity to the Sun. We must either view Mercury during broad daylight or when it is low on the horizon before sunrise or after sunset. During these latter times, we are viewing through thick layers of unstable air.

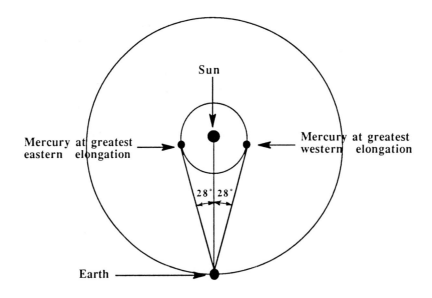

Figure 3.1: Elongation of Mercury

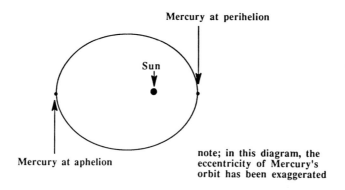

Figure 3.2: The Orbit of Mercury

Nevertheless, at the end of the last century, attempts were made to determine the rotation period of Mercury by "tracking" faint surface features as the planet turned. These observations, however, were consistent with several possible rotation periods; among them 44d, 59d and 88d. Of these possibilities, the 88d rotation period was chosen as correct, not because it was observationally more likely, but because it was believed that Mercury should be in the same sort of synchronous* relationship with the Sun as the Moon is with the Earth. As it turned out, this was wrong.

In the mid-1960's, astronomers started bouncing radio waves off the nearer planets. By receiving the reflected signals, much information could be deduced about these planets; their distance from the Earth, their surface features and their rotation periods. When this was done for Mercury, it was discovered that its rotation period was 59d, not the 88d that had been believed for 75 years.

*A "synchronous" rotation is one where the rotation and revolution periods are the same so that the orbiting body keeps the same "face" to the central body.

At first glance, the 59d rotation period of Mercury seems particularly uninteresting, but there is more to it than immediately meets the eye. It turns out that 59 = 2/3 (88)! This means that after 1 sidereal rotation period (59d), Mercury has moved 2/3 of the way around its orbit, or that in 1 sidereal revolution period (88d) Mercury has rotated 3/2 times.

This phenomenon is depicted in Figure 3.3. Assume that a pole is buried in the soil of Mercury and that it is pointing straight "up". Further assume that the pole is pointing towards the Sun when the planet is at perihelion. Figure 3.3 illustrates that since Mercury rotates 3/2 (= 1 1/2) times per revolution, when perihelion is again reached (88d later), the pole will be pointing directly away from the Sun. After another 88d period, the pole will again be pointing at the Sun and one solar day on Mercury will be over [The solar day on Mercury is 176 Earth days long, while its sidereal day (period of rotation with respect to the stars) is 59d long*].

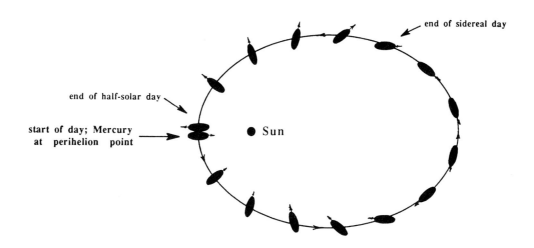

Figure 3.3: Rotation of Mercury

The reason for this phenomenon is that there is a gravitationally induced "bulge" on the surface of Mercury which has been "captured" by the Sun. The "capture", however, is not as strong as in the Earth-Moon system where the Earth has forced the bulge on the Moon to perpetually point towards the Earth. In the Mercury case, the "bulge" is probably an elongated axis (as indicated in the last diagram) and the effect of the Sun is to force alternate ends of the long axis to point towards the Sun at successive perihelion passages (point of maximum solar gravitational effect). The Mercury case is known as a <u>1 1/2:1 spin-orbit couple</u>.

An interesting effect occurs as Mercury orbits the Sun. For an observer on Mercury, the slow eastward rotation causes the Sun to slowly move in the sky from east to west. The effect of the eastward motion of the planet about the Sun causes the Sun to slowly drift eastward in the sky, but this effect is less than the westward solar drift due to rotation. (This is the same on the Earth where the Sun moves 360° each day towards the west because of rotation and 1° each day towards the east because of revolution.)

On Mercury, however, the length of the day and the length of the year are not that different. During the time that Mercury is near perihelion, it speeds up in its orbit and this hastens the Sun's apparent eastward motion to the point where it, for a few days, overtakes and surpasses the Sun's apparent westward motion due to rotation. Then net effect is that for most of the time the Sun will

*For a discussion of solar and sidereal periods see page 96.

move slowly from east to west across the Mercurian sky. Then this motion will slow and reverse and the Sun will start moving backwards. This will continue for a few days until the direction rights itself. At certain places on Mercury, the Sun will rise in the east, come to a stop in the sky, turn around and immediately go to its setting in the east only to rise again a few dozen hours later.

Observations of Mercury; Detailed observations of Mercury are difficult from the surface of the Earth for reasons already discussed but certain properties of the planet are readily ascertained.

From the known distance and angular size of Mercury, its diameter can be shown to be about 3000 miles; less than half that of the Earth (Figure 3.4).

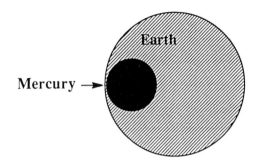

Figure 3.4: **The Diameter of Mercury**

From its gravitational effect on passing comets and space probes, the mass of the planet is known to be about 1/20 (5%) that of the Earth.

From the known size and mass of Mercury, its density (mass/volume) can be calculated to be 5.5 grams/cc. This is the same density as the Earth and is the highest in the Solar System. Since Mercury is less massive than the Earth, it has less gravity and is less gravitationally compressed. To have the same mean density, it must contain a higher proportion of heavy materials, principally iron. It is suspected that 60 percent of the mass of Mercury is comprised in its nickel-iron core.

From the brightness of the planet and its distance from the Sun, we can compute the fraction of incident sunlight reflected by Mercury (this is called **albedo**). The albedo of Mercury is only 6 percent. Since the albedo of the Moon is 7 percent, one might hypothesize similar surfaces. Since Mercury is close to the Sun (hot) and of relatively small mass (low gravity), we might assume that it would have little or no atmosphere and that there would be no liquid water. This also argues for a lunar-type surface.

Even though certain planetary properties can be deduced from the Earth, there is no substitute for "being there". Therefore, in 1974 the United States sent the Mariner 10 spaceprobe on a flyby mission to Mercury. This probe transmitted, back to Earth, views of the Mercurian surface and data concerning its surface properties.

Mariner 10 was put into an orbit about the Sun which allowed it to closely view Mercury three separate times (Figure 3.5). In all, over 4000 images of Mercury's surface were transmitted to Earth. Two of the best views are shown in Figure 3.6 and 3.7. As is clearly seen, the surface of Mercury strongly resembles the surface of the Moon. Thousands of craters are seen ranging in size up to more than 100 miles in diameter. Basins (empty "seas") up to almost 100 miles in diameter are also seen. Cliffs (called "scarps") almost a mile high and hundreds of miles long are in abundance. It is believed that these scarps are the result of the shrinking of the large iron core of the planet and the subsequent "wrinkling" of its crust (the way the skin of a prune wrinkles as the fruit shrinks while drying). Some craters show a system of rays* emanating from them. These rays do not seem to be associated with the youngest craters, as they are with the Moon. Also, Mercurial rays do not extend as far as those found on the Moon. This is due to the higher gravity on Mercury.

*Rays are bright streaks emanating from young craters. The same phenomenon is seen on the Moon (Chapter 5).

Because of the higher surface gravity, the crater rims on the surface of Mercury have a lower elevation than rims of lunar craters of the same diameter. As with the Moon, most of the craters are formed from impacts with asteroids and comet nuclei.

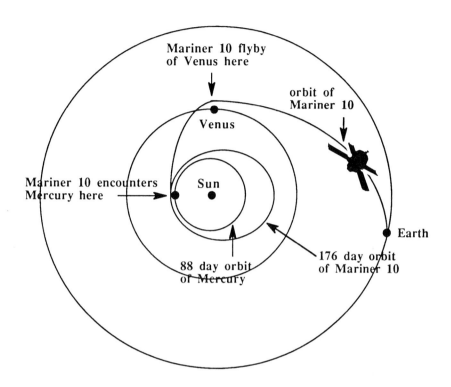

Figure 3.5: The Orbit of Mariner 10

While craters on the Moon are named after famous scientists, on Mercury the craters are named after noted personages from the arts (e.g. Bach, Beethoven, Mozart, Matisse). The scarps are named for ships of discovery (e.g. Santa Maria) and the plains are named for namesakes of the god Mercury (e.g. Odin).

Infrared radiometers (devices for measuring the amount of infrared radiation emanating from the planet) aboard Mariner 10 indicated noontime temperatures exceeding 600°F dropping to almost -300°F at night. This is the greatest day/night temperature swing in the solar system. The rate at which the temperature dropped after sunset on the Mercurian surface indicated a layer of insulating dust several inches thick.

An ultraviolet spectrometer aboard Mariner 10 indicated the presence of an extremely thin helium atmosphere (about a billionth as dense as Earth's atmosphere). This gas is thought to be captured from the solar wind or outgassed from radioactive decay in the crust of the planet.

Recent ground-based observations have shown that the most abundant element in the tenuous atmosphere of Mercury is sodium; accounting for 50 times more material than helium. It is believed that the atoms of sodium are ejected into the atmosphere of Mercury as particles from space (cosmic rays and micrometeorites) impact the surface of the planet; this process is called "sputtering".

The most unexpected discovery of the Mariner 10 mission was that of a magnetic field surrounding the planet. The field is weak (only about 1 percent as strong as Earth's) but stronger than one would expect from so slowly rotating a planet or one of too low a mass to have a molten core. Perhaps the field was "frozen" into the core at the time of its formation.

Figure 3.6: Mariner 10 View of Mercury

Mercury and Relativity; Early observations of the orbit of Mercury indicated that its perihelion point was slowly precessing about the Sun (Figure 3.8). The magnitude of this advance is 5600 arc seconds per century (just over 1 1/2 degrees). According to Newton's laws of motion and gravitation, the bulk of this effect can be explained as the result of the gravitational effect of the planets in the Mercury-Sun orbit. However, Newton's laws only predict a perihelion advance of 5557 arc seconds per century. How do we explain the observed additional 43"/century? The observations are too easily made to be incorrect.

The answer to the question came in 1916 when Albert Einstein published his **General Theory of Relativity**. One of the predictions of this theory of gravity is that in the vicinity of the Sun, time and space are curved. The effect of this is to make bodies move in a slightly eccentric manner when compared to Newtonian theory. When calculated precisely, the effect of relativity on the orbit of Mercury is exactly 43.06 arc seconds per century; exactly what is observed. This observation has been used as one of the confirmations of the General Theory.

Figure 3.7: Mariner 10 View of Mercury

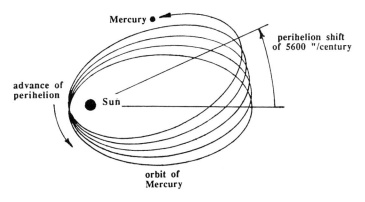

Figure 3.8: Advance of the Perihelion Point of Mercury

VENUS

Sometimes known as the Earth's "sister" planet, Venus has a size, mass and density which are almost the same as its twin. Furthermore, Venus is the closest planet to Earth, sometimes getting as close as 25 million miles. Add to this the observation that Venus has clouds in its atmosphere, and one sees why observers might believe that Earth and Venus could be quite similar.

Figure 3.9(a): Venus from Mariner 10

Figure 3.9(b) Venus from Magellan

Venus is 0.7 au from the Sun (about 65 million miles) and orbits it in 225 days (Learning Activity 8, Appendix IV). Because of its proximity to the Earth and its white/yellow highly reflective cloud cover, Venus is the 3rd brightest object in our sky; out-shined only by the Sun and the Moon. As with Mercury, Venus is sometimes seen after sunset in the evening sky and sometimes before sunrise in the morning sky. The early Greeks thought it was two objects and called it "Phosphorus" when it was in the morning sky and "Hesperus" when it was an evening object. The maximum elongation of Venus is 47°.

As has already been stated, Venus goes through a series of phases. This was first observed by Galileo in 1610. Contrary to what one might expect, Venus is not its most brilliant when it is at full phase because at this time it is farthest from the Earth. Nor is it its most brilliant when it is nearest the Earth because it is then at new phase. Venus is most brilliant when it is a crescent phase at an elongation of 39°. This occurs 36 days before and after inferior conjunction (the time of closest approach to Earth). As with Mercury, Venus has no known satellites.

Rotation of Venus; Because of its dense cloud cover, the surface of Venus cannot be seen from the surface of the Earth. It is, therefore, not possible to watch surface features as they rotate with the planet.

In 1962, radio (radar) waves were bounced off the surface of Venus. These long wavelength radiations have no difficulty penetrating the thick Venusian cloud layers. The returned radiation was shifted in wavelength by the rotation of the planet. From these studies, it was determined that Venus rotates "backwards" (retrograde) with a period of 243 days. Venus is the only planet in the solar system which rotates clockwise* (towards the west). It also has the slowest rotation rate.

The rotation and revolution period of Venus conspire to face the same hemisphere of Venus towards the Earth at each inferior conjunction (It is not known if this is a coincidence or a form of spin/orbit coupling.).

The Atmosphere of Venus; Spectroscopic studies from the surface of the Earth have indicated that 96 percent of the Venusian atmosphere is composed of carbon dioxide. The rest of the atmosphere made up of nitrogen and argon. This is in stark contrast to the atmosphere of the Earth which is 78 percent nitrogen and 21 percent oxygen. The white/yellow clouds of Venus have been shown to be made of droplets of sulfuric acid in contrast to Earth clouds which are composed of drops of condensed water.

In 1978, the Pioneer Venus Probe discovered that the sulfuric acid clouds of Venus do not extend lower than about 30 miles above the planet's surface (nor higher than about 36 miles). The reason for this is that below this level, the temperature is too high for drops of sulfuric acid to condense.

The Temperature and Pressure at the Venusian Surface; If one knows the distance of a planet from the Sun, one can calculate how hot it should be at its surface. Using this reasoning, scientists calculated that the temperature at the surface of Venus should be about 140°F. In the 1960's, however, infrared sensors measured the quantity of long infrared radiation leaving the surface of Venus. The amount of radiation was consistent with a surface temperature of almost 900°F! Subsequent measurements by Venus landers have confirmed this number. Venus has the hottest surface of any planet in the solar system. Even the night side of Venus is at a temperature of 850°F. Why is this planet so hot?

The answer to this question is known as the **greenhouse effect**. Visible radiation from the Sun penetrates the clouds of Venus and is absorbed by the surface, heating it up. The hot surface re-radiates the energy it absorbs as infrared radiation. But the atmosphere of the planet (CO_2), while transparent to visible light, is opaque to infrared radiation. The radiation is, therefore, trapped by the atmosphere and not allowed back into space. The surface of the planet gets hotter and hotter until the amount of radiation leaving the top of the atmosphere just matches the amount of radiation entering. This match occurs at a surface temperature of 900°F (Figure 3.10).

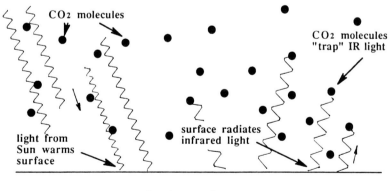

Figure 3.10: The Greenhouse Effect

*Uranus and Pluto have rotation axes which are highly tilted and, thus, their directions of rotation are ambiguous.

The high temperatures on Venus are a result of the CO_2 atmosphere. Why has this not happened to Earth? The reason is that the Earth has little CO_2 in its atmosphere. This was not always the case. The early atmosphere of the Earth, outgased from volcanoes, and perhaps collected from collisions with comets, was mostly carbon dioxide and water. At the Earth's surface, however, the temperature was low enough to allow water to condense into oceans. The CO_2 in the atmosphere then dissolved in the oceans (forming carbonated water) and precipitated out with other minerals as calcium carbonate rocks (limestone). The CO_2, formerly in our atmosphere, is now in our sedimentary rocks.

On Venus, it was too hot for the outgassed water vapor to condense into oceans (In fact, it is believed that the water vapor was dissociated by ultraviolet radiation into hydrogen and oxygen. The hydrogen, being light, escaped into space and the oxygen combined with minerals on the Venusian surface or with molecules in the Venusian atmosphere). The CO_2, therefore, remained in the atmosphere where it today is responsible for the high surface temperatures.

The weight of the CO_2 in the atmosphere is very great. This causes a high atmospheric pressure at the surface of the planet. The weight of the Earth's atmosphere is 14.7 pounds per square inch at the Earth's surface. On Venus, the weight of the atmosphere is 90 times as great (1300 pounds per square inch). This is equivalent to the pressure over 1/2 mile below the surface of one of Earth's oceans.

Space Probes to Venus; The first spacecraft to successfully land on Venus was the Soviet Venera 7 lander (1970) which transmitted data for 23 minutes from the Venusian surface. In 1972 Venera 8 transmitted for 50 minutes before going mute. Both probes confirmed Earth-based findings of Venusian surface conditions.

In 1974, on its way to Mercury, Mariner 10 took thousands of photographs of the cloud tops of Venus. One of these is shown in Figure 3.9. These Mariner 10 photos indicate that while the body of the planet rotates once every 243d, the upper atmosphere circulates once every 4d and is, therefore, the Venusian counterpart of the Earth's jet stream. Mariner 10 detected no magnetic field; an indication that either no liquid iron core exists or that the rotation rate of the planet is too slow to induce an Earth-type field.

In the years after 1974, both the Soviet Union and the United States sent additional landers to Venus. The later Soviet Venera landers were equipped with cameras and floodlights to photograph the landing areas (the floodlights were not needed as more light penetrates the cloud layers than was suspected). (Figure 3.11)

Some Venera landing sites showed sharp, angular rocks while others showed flat, smooth, weathered rocks. Several areas showed what looked like cooled lava. Since the measured wind speed at the surface of Venus is only 1 to 2 miles/hour, it is unlikely that the weathering of the rocks is due to wind (as in the "red rock" country of Utah and Arizona). It is more likely that the weathering is due to melting or chemical changes.

In 1983, Veneras 15 and 16 went into orbit around Venus and carried out surface mapping of the northern hemisphere using radio waves to penetrate the clouds and "illuminate" the surface.

In 1978, the United States sent its Pioneer Venus multiprobe to Venus. The multiprobe deployed five separate packages into the Venusian atmosphere; a large probe, 3 small probes and the "bus"; the vehicle on which the probes were mounted. The large probe descended on a parachute while the small probes fell freely. Despite this, one of the small probes survived the impact and transmitted data for over an hour. All of the probes sent to Venus have confirmed and reconfirmed the properties already stated.

Surface of Venus; Despite its dense cloud cover, the surface of Venus has been mapped from the Earth and from space probes by bouncing radar (radio waves) off the surface. In 1978, the Pioneer Venus radar mapper probe mapped all but he polar regions of Venus. This revolutionary global picture is shown in Figure 3.12.

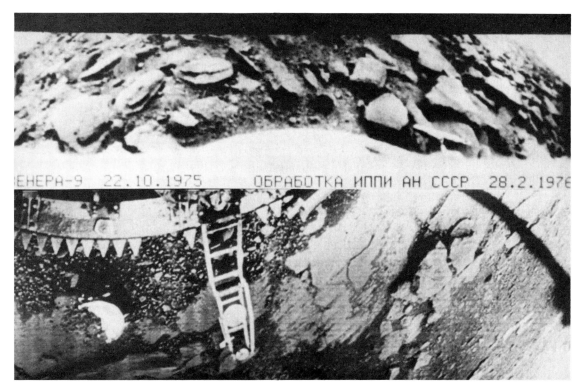

Figure 3.11: Venera Photos from the Surface of Venus

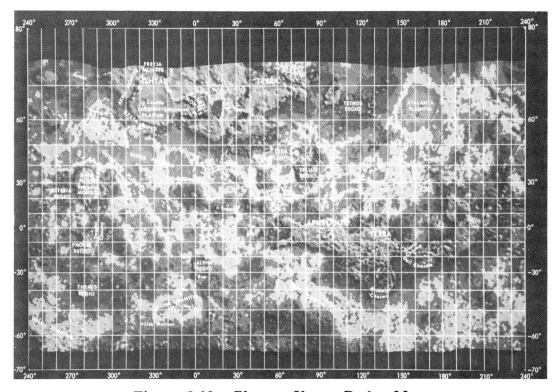

Figure 3.12: Pioneer Venus Radar Map

As with the Earth, Venus shows low regions (similar to our ocean basins) and high areas (similar to our continents). The lowland plains comprise 60 percent of the surface of Venus. There are two continents on Venus similar to Australia in size and rising one to three miles above the lowlands.

Soviet radar mapping probes have discovered volcanic mountains with distinct calderas at their summits. One such mountain, called Beta, is 500 miles in radius with a crater at its peak almost 30 miles in diameter. Many examples of such volcanic peaks exist.

On the flat Venusian plains are clear examples of impact craters, some as large as 120 miles in diameter. From the number of these craters, it is possible to infer that the average age of the Venusian surface is of the order of 1/2 billion years; intermediate between the ages of the Lunar and Terrestrial surfaces.

Most surface features on Venus have been named after mythological female personages (e.g. Aphrodite, Ishtar, Eve) a notable exception being the Maxwell mountains named after the famous physicist James Clerk Maxwell.

Figure 3.13: Venusian Lava Flow

In 1990 and 1991 the Magellan Radar Mapper, in orbit about Venus, mapped the entire Venusian surface in strips 10 x 17 miles wide. Composite photos, made from these strips, have revealed a strange and chaotic surface and have given us new insight into the past and present of our nearest neighbor (Figure 3.9b).

Images from Magellan confirm that volcanism is widespread on Venus. Four fifths of the planet's surface is covered by large lava flows of basalt type rock (Figure 3.13). Also observed are thousands of small volcanoes, volcanic craters (Figure 3.14), and pancake domes (Figure 3.15). Further evidence for volcanic activity comes from the detection of sulfur dioxide in the Venusian atmosphere. This gas is emitted by volcanoes and is continually dissociated by ultraviolet light. Its presence, therefore, is circumstantial evidence that volcanic outgassing presently occurs.

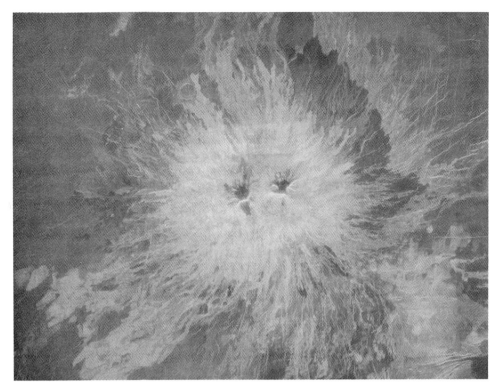

Figure 3.14: Sapas Mons Volcano (250 miles x 1 mile high)

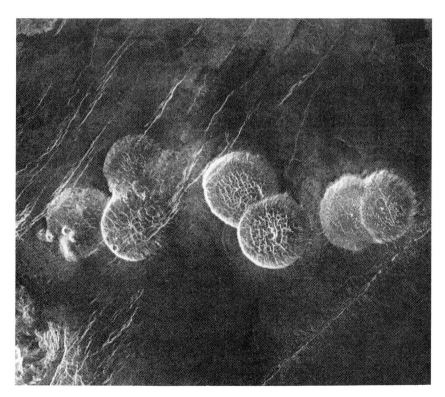

Figure 3.15: 15 Mile Diameter Pancake Domes

Lava channels, "riverbeds" cut by flowing lava, are found running for tens to hundreds of miles (Figure 3.16).

Impact craters are common on the Venusian surface but large areas <u>without</u> craters indicate the volcanism has been locally efficient. The lack of craters less than two miles in diameter indicates that small objects disintegrate upon entering the Venusian atmosphere (Figure 3.17).

Mountain building on Venus apparently works in a manner which is different from that which occurs on Earth where the collisions between crustal plates cause the plate edges to buckle and rise vertically. All available evidence indicates that Venus <u>does not</u> have continental drift. How then are mountains, such as Maxwell Montes, and highlands, such as Ishtar Terra (Figure 3.18) formed?

Evidence now indicates that the mountains and uplifted areas on Venus may be due to hot material rising to the surface and producing mountains (similar to our Hawaiian Islands). Similarly, hot materials trapped under the crust might begin to flow horizontally, dragging crustal material with it. This dragged material would "pile up" and "fold", creating uplifted and mountainous terrain. In any case, it appears that the Venusian highlands are the result of volcanic activity rather than continental drift.

Figure 3.16: Lava Channels on Venus

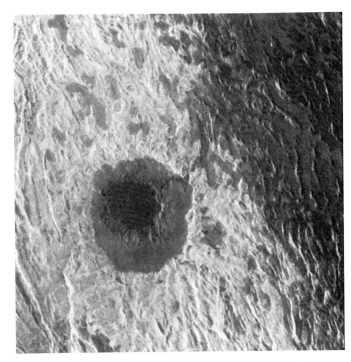

Figure 3.17: Impact Crater "Cleopatra" (1 1/2 miles deep)

Figure 3.18: "3D" View of Western Ishtar Terra

MARS

In 1869, Father Angelo Secci, while observing Mars, saw faint linear features on the planet's surface. He called these features "canali" which in Italian, means "channel" or "groove".

In 1877, the Italian astronomer Giovanni Sciaparelli published the results of many years spent observing **The Red Planet**. His observations showed that Mars had, on its surface, dark areas which Sciaparelli suspected might be bodies of water. Some of these dark areas appeared to be connected with the narrow dark markings observed by Secci. He thought that these narrow areas could be waterways connecting the larger seas (much as the English Channel connects the Atlantic and the North Sea). Sciaparelli, therefore, concluded that the "canali" on Mars might be liquid in nature (Figure 3.19).

Figure 3.19: Early Map of Mars

When Sciaparelli's work was translated into English, the word "canali" was translated as "canal" rather than as "channel", and the word canal connotes an artificial construction. In the United States, **Percival Lowell**, a wealthy Bostonian and amateur astronomer, became interested in the "canals" of Mars. In 1894, he built an observatory at Flagstaff, Arizona dedicated to the study of Mars and its canals.

Lowell was convinced that Mars was populated by an advanced race of intelligent creatures and that they were engaged in a life or death struggle to save a dying planet. Mars, he thought, was losing its atmosphere and its water; both of which were evaporating into space. Lowell believed that the only water left on Mars was frozen into the polar caps, which were easily observed from Earth. The Martians, however, had to live near the equator where the temperatures were more moderate. In order to "get the water", the Martians had engineered a vast network of canals which could be seen from Earth-based telescopes. This was Lowell's concept and his writings and lectures initiated a popular myth which remains with us today. One of Lowell's maps is shown in Figure 3.20, and a photo of Mars is shown in Figure 3.21.

Modern observations (including photos from Mars probes) indicate that there are no canals on Mars and that Lowell's maps do not match with any known Martian feature. Lowell's Martians apparently do not exist, but Lowell's interest and enthusiasm led to an enhanced public awareness in Mars in particular and in astronomy in general. This interest ultimately led to our explorations of our mysterious neighbor.

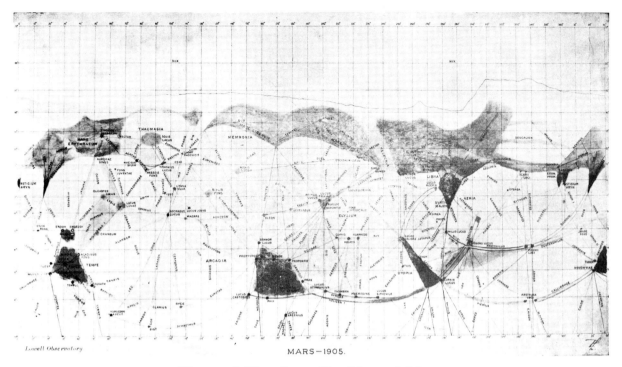

Figure 3.20: Lowell's Map of Mars

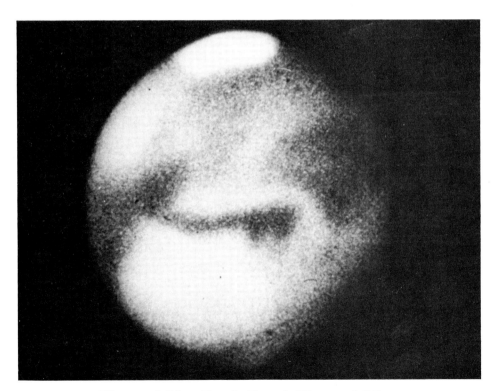

Figure 3.21: Earth-Based Photo of Mars

Appearance of Mars; With the naked eye, Mars appears as a ruddy, yellowish, starlike object. Small aperture telescopes will not improve this very much. In an 8" telescope, one might notice that either the top or the bottom of the planet appears whitish. These white areas are the Martian polar caps (Figures 3.21, 3.22). These caps alternate in size. When it is winter in the Martian northern hemisphere, the northern cap is large. At that time it is summer in the southern hemisphere and, hence, the southern cap is smaller or non-existant. The reverse is true when it is summer in the northern hemisphere. The reason for the Martian seasons is its 25° axial tilt.

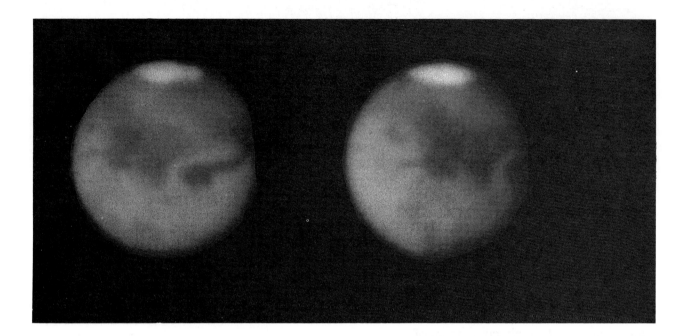

Figure 3.22: The Martian Polar Cap

Also noticable through a small telescope are fuzzy dark markings.* It is these markings that some early astronomers believed to be bodies of water. Other astronomers, however, noticed that these dark markings seemed to fade in the Martian winter and they hypothesized that they may actually be forests or some form of vegetation which grew and died with the seasons. These ideas lent credence to Lowell's "Martians" (Figure 3.23).

Properties of Mars; Mars orbits the Sun in just about 2 Earth years (more precisely, 23 months). Its period of rotation (determined, by Cassini in 1665, by watching surface features as the planet turns) is 24h 37m. The fact that this number is close to the figure for the rotation period of the Earth is coincidental; the rotation period of the Earth is slowly changing due to lunar interactions. In the future, our rotation period will be much longer than 24h; in the past it was much less.

The diameter of Mars is just over 4000 miles or about half the diameter of the Earth. The mass of the planet is about 1/10 the mass of the Earth. This number is determined by observing the gravitational effect of Mars on its satellites (Learning Activity 9, Appendix IV). From the size and mass of the planet, we can calculate the density and surface gravity of Mars. Its density turns out to be 4 grams/cc (similar to the Moon) and its surface gravity is 38% that of the Earth. Mars is

---

*Through a small telescope, the best observations of Mars are made when Earth and Mars are at their closest. This occurs when both Earth and Mars are on the same side of the Sun; a condition known as **opposition** (because Mars and the Sun are opposite one another in sky). Oppositions of Mars occur at intervals of two years and two months.

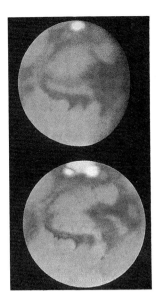

Figure 3.23: The "Changing Colors" of Mars

about 140 million miles from the Sun (about 1 1/2 au), and, at its closest, is 35 million miles from the Earth. Mars is the 2nd closest planet to the Earth.

Exploring Mars; From the surface of the Earth, it is difficult to learn much about Mars beyond its gross properties. Spectroscopic work has yielded some basic information about its atmosphere. The main component of the Martian atmosphere is carbon dioxide, comprising some 95%. Three percent of the atmosphere is nitrogen and two percent is argon. The atmosphere of Mars is similar to that of Venus in composition but very different in pressure. While the pressure at the surface of Venus is 90 times that at the surface of the Earth, the pressure of the atmosphere of Mars in only one percent that of Earth's.

In 1965, the Mariner 4 spacecraft flew by Mars and radioed back to Earth 22 photos of the Martian surface. These photos were extremely disappointing. They showed a bleak, cratered world devoid of interesting features. The best of these is shown in Figure 3.24.

In 1969, Mariners 6 and 7 confirmed that Mars was cratered. No evidence was found for any internal activity (volcanoes) or mountains. It seemed to be a Moonlike world covered by a thin carbon dioxide atmosphere. Mars was boring!

The irony of the situation is that Mars has an interesting side and a mundane side. This was not discovered until 1971 when the United States put Mariner 9 into orbit about Mars. From this vantage point, the spacecraft could photograph the whole Martian surface in great detail. In all, Mariner 9 discovered volcanoes, deserts, canyons, channels apparently cut by running water and intricate polar layering. Mars was a real planet! The first craft to "touch" Mars was Mars 2, a Soviet "lander" which crashed on the planet in 1971. In the same year, Mars 3 successfully landed and transmitted data for 20 seconds before it fell forever silent.

In 1976, the United States sent two probes, Viking 1 and Viking 2, to Mars. Each probe consisted of an orbiter and a lander. On July 20, 1976, the Viking 1 lander touched down on the surface of **Chryse Planitia** (the Plains of Gold). In September of the same year, the Viking 2 lander landed on the surface of **Utopia Planitia** (Figure 3.26). These two landers contained cameras, soil test equipment, weather stations and seismic stations. While the landers were testing the Martian surface, the orbiters were photographing the planet below. Much of what we know about Mars comes from the Viking missions. The Viking 1 lander transmitted from the surface of Mars for six years.

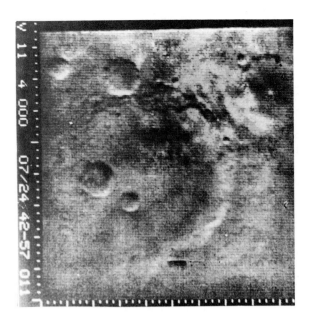

Figure 3.24: Mariner 4 Photo of Mars

The Surface of Mars; The Viking landing sites (Figure 3.25) turned out to be desert-like areas covered with reddish soil and boulders up to ten feet in diameter. The soil composition is high in iron oxide, which is why it appears red. The Martian sky appears pink because of suspended dust particles scattering red light. The winds at the landing sites averaged 10 to 30 miles per hour (rising to 200 mph during storms). The temperature at the landing sites varied from -20 °F to -120 °F (At the poles, the temperature drops to -200 °F and at the equator it rises to +70 °F).

The Volcanoes of Mars; The surface of Mars is roughly divided into two major types. Most of the southern hemisphere consists of ancient cratered terrain, analogous to the Moon and Mercury (although not as heavily cratered). Most of the northern hemisphere, however, is made up of younger volcanic plains. The volcanic plains are, on the average, several miles lower than the cratered terrain. The process which destroyed the cratered terrain and lowered the plains is unknown. (Figure 3.27)

The largest features in the southern "highlands" are ancient impact basins. The largest of these basins, **Hellas**, is over 1000 miles in diameter.

On one side of the planet is a "North American" size bulge rising six miles above the surrounding plains; the **Tharsis bulge**. On this bulge are found the four great volcanoes of Mars (Figure 3.29). The three volcanoes on the right are named (N to S) **Mt. Ascraeus, Mt. Pavonis** and **Mt. Arsia**.

Of these four volcanoes, the largest, **Mt. Olympus**, has a base which is 300 miles in diameter and a height of 15 miles. The caldera at the top of the volcano is almost 50 miles across. This volcano is larger than any other such feature found in the solar system and nearly twice as high as Mauna Loa, the largest volcano on Earth. (Figures 3.30 and 3.31)

In all, about a dozen large, and many smaller volcanoes have been found on Mars. The large volcanoes are generally associated with the Tharsis region or its environs. From the number of impact craters in its vicinity, it has been determined that Mt. Olympus is rather young; probably less than 1 billion years old. It is even possible that it is still active. Further evidence for the youth of this feature comes from the "fresh" looking lava flows on its flanks. The reason that the Martian volcanoes are larger than Earth volcanoes is that Mars does not have continental drift to carry forming volcanoes away from "hot spots" in the planetary mantle. They, therefore, keep on building to high elevations.

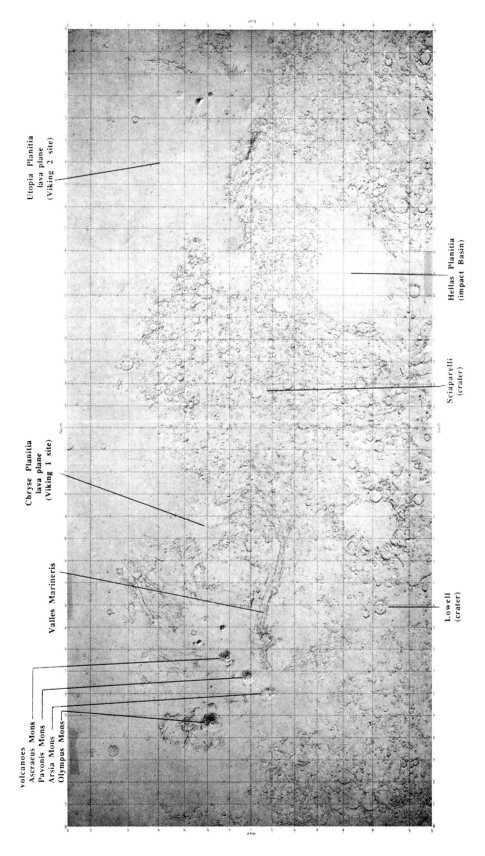

Figure 3.25: The Viking Landing Sites

Figure 3.26: Mars as seen from the Viking Lander

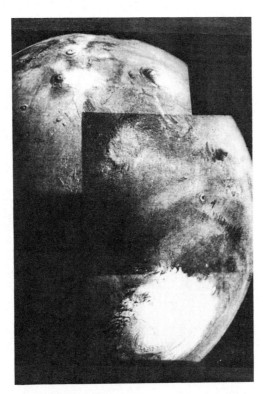

Figure 3.27: Northern and Southern Hemispheres of Mars

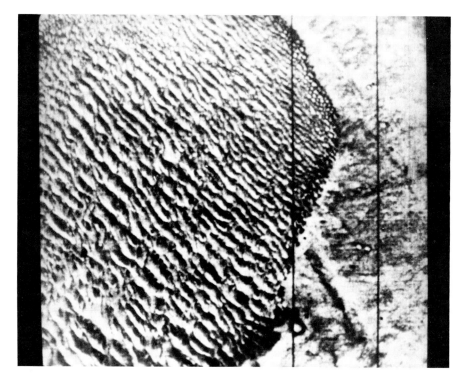

Figure 3.28: Sand Dunes on Mars

Figure 3.29: The Tharsis Bulge

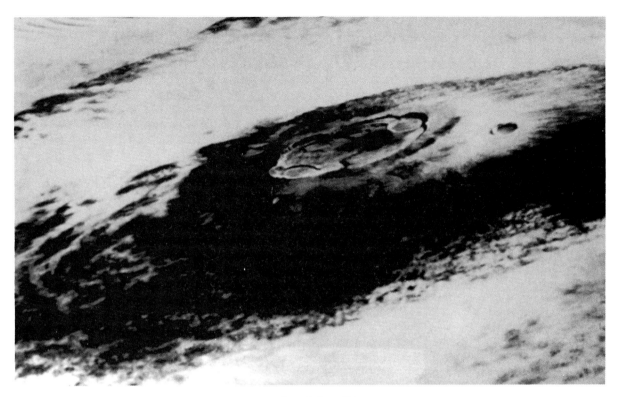

Figure 3.30: Mt. Olympus

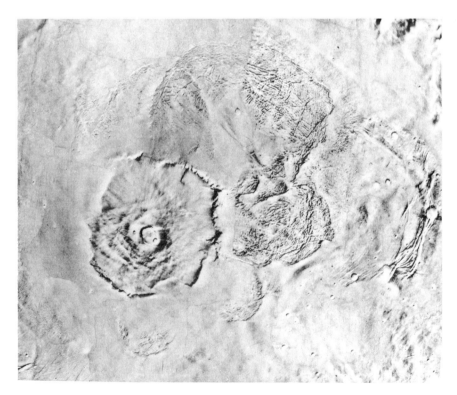

Figure 3.31: Mt. Olympus

The Canyon of Mars; The largest canyon in the solar system (so far!) is a 3,000 mile long, 3.7 mile deep, 60 mile wide gash located just east of the Tharsis volcanoes. The feature, named **"Valles Marineris"** (Valley of the Mariner), is a giant crack in the surface of Mars caused by internal pressures; the same pressures that formed the Tharsis features to the west.

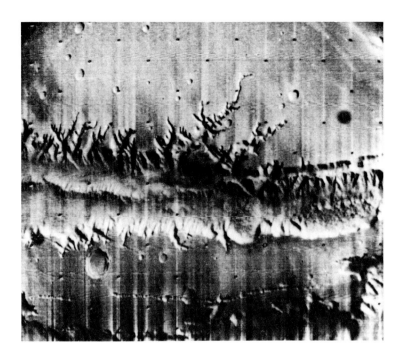

Figure 3.32: Valles Marineris

Both the Tharsis bulge and Valles Marineris are thought to be 2 to 3 billion years old. In the time since the formation, the canyon apparently had been widened and sculpted by winds and, perhaps, running water (Figures 3.32, 3.33, and 3.34).

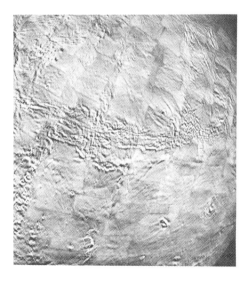

Figure 3.33: Valles Marineris

Figure 3.34: Valles Marineris

Water on Mars; Aside from the water frozen in the polar caps (discussed later), there is evidence that large quantities of water once flowed freely on the Martian surface. This evidence comes from Viking orbiter photos of surface channels.

One type of channel, called **runoff channel**, is a few yards deep and about 10 yards wide. They extend for 5 to 10 miles and probably carried runoff during ancient rainstorms (there is not enough atmospheric water today to allow rain to form). They are generally found on the ancient cratered uplands. We can tell when these channels were formed by determining the age of the surface on which the channels are found. This is accomplished by doing statistical studies of the numbers of craters of various sizes. The time of the last rainstorms seems to be about 4 billion years ago (Figure 3.35).

Another line of evidence for Martian surface water comes from huge **outflow channels** that connect the ancient Martian uplands to the northern volcanic plains. These channels, which are up to 10 miles wide and hundreds of miles long, are too large to have been formed by rainfall and are probably the result of catastrophic or "flash" flooding. The source of the water for such flooding is unknown but might be the sudden melting of large quantities of water from a permafrost layer. Such melting could be connected to volcanic activity such as the formation of the volcanic plains. (Figure 3.36).

For such quantities of liquid water to have existed, Mars must have once had a dense atmosphere, perhaps as dense as Earth's. Estimates indicate that Mars may once have had enough surface water to make an ocean 20 feet deep over the entire planet! Is the water gone? Maybe not. It could be frozen into a permafrost layer and into the Martian polar caps. The current "ice age" may have been triggered by a change in the axial tilt of the planet. This could have been accomplished by a redistribution of mass during volcanic outbursts.

Figure 3.35: Runoff Channels

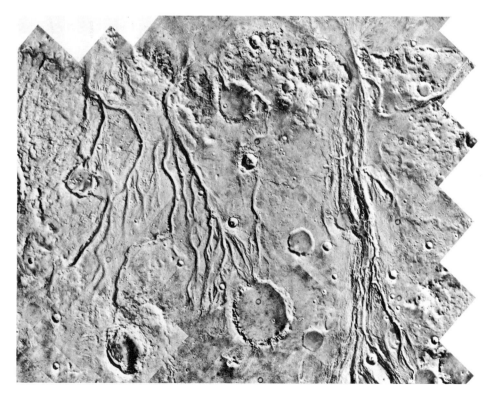

Figure 3.36: Outflow Channels

The Polar Caps; The Martian polar caps, easily seen through Earth-based telescopes, have been extensively photographed and studied by orbiting satellites. Unlike Earth's permanent polar caps, the Martian caps are composed of frozen water <u>and</u> frozen carbon dioxide (dry ice). When the surface temperature drops below -190 °F, atmospheric CO_2 freezes directly onto the surface. These cold temperatures are reached in the alternating southern and northern polar winters. The southern polar cap is larger and more defined than the northern cap since Mars is farther from the Sun during the southern winter than during the northern winter. These "seasonal" caps grow and shrink with the seasons and are only a few feet thick.

Underneath the seasonal cap are the "permanent" caps (which may be up to two miles thick) which remain through the polar summers. The permanent southern cap is composed of H_2O and CO_2 while the permanent northern cap is composed of frozen water only. This is perhaps explained by the fact that the Martian seasonal dust storms take place during the northern summer. Dust might then cover the northern cap, darkening it so that it absorbs solar radiation and warms to the point where CO_2 will vaporize.

The polar terrain, exposed when the ice recedes, is quite unusual. At latitudes above 80°, the surface consists of a series of sedimentary plates. Each plate is about 30 feet thick and may be composed of compacted equatorial dust, carried to the polar regions by the seasonal dust storms. It is believed that each plate might represent the collected deposits of 10,000 years worth of dust storms (Figure 3.37).

The seasonal dust storms also explain the "seasonal changes" observed by Lowell and others. Areas of dark rocks, alternately exposed and covered by dust were mistakenly interpreted as vegetation growing during the Martian spring and dying during the Martian fall and winter.

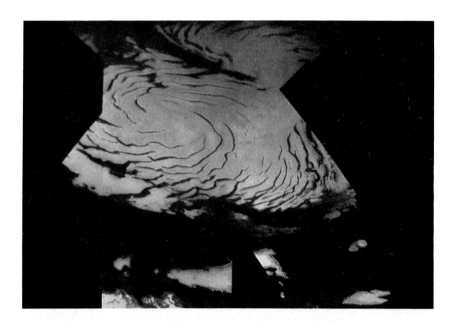

Figure 3.37: Sedimentary "Plates" at South Pole

The Viking Experiments; Our probes of the planet Mars have answered many long-standing questions about our mysterious neighbor. Lowell's canals have turned out to be non-existent. The changing colors of the dark areas, which some thought to be evidence of plants, have turned out to be dark landforms covered and uncovered by seasonal dust storms. The "ice caps" of Mars have been discovered out to be partially "dry-ice caps". The more we look and study, the less likely it seems that life will be found on Mars. Since the landing of Viking, the hope has practically dissolved.

The Viking landers at Chryse and Utopia have found a barren, desolate, boulder strewn, stark landscape, showing no evidence of life. Aside from the TV cameras, the Viking landers carried weather stations, seismic stations and a sophisticated biological laboratory. This "life detection test" consisted of 3 separate experiments. Each experiment was supplied with a Martian soil sample from a small mechanical scoop which gathered a small quantity of surface material and distributed it to the three separate chambers.

- **The Gas Exchange Experiment:** The Martian soil is mixed with a small amount of water and other nutrients. If micro-organisms exist in the soil and if they metabolize the water or nutrients, the atmosphere in the chamber should change composition as the organisms expel gases.

- **The Labeled Release Experiment:** The nutrients which are fed to the Martian soil are "tagged" with radioactive atoms. If micro-organisms metabolize the nutrients and then expel waste products, some of the radioactive atoms should find their way into the atmosphere of the chamber. Detection of radioactive atoms in the gases above the soil would be evidence for life.

- **The Pyrolitic Release Experiment:** Martian soil is mixed with "tagged" CO_2. If there are micro-organisms, and if they metabolize the CO_2, then the radioactive atoms might become part of their bodies. The CO_2 is then evacuated from the chamber and the soil is analyzed. Any radioactivity remaining would be evidence of microbial action.

Each of the three life-detection tests initially gave positive results at both landing sites. As time went by, however, the results became weaker and weaker as if the organisms were becoming less active. This is not the way terrestrial organisms would behave.

A 4th experiment was performed which makes the results of the first 3 all the more puzzling. The apparatus is called the "gas chromatograph mass spectrometer (GCMS)" and its function was to analyze the composition of the Martian soil. The strange result was the total absence of organic material; chemicals which are the stuff of which Earth-life is composed. How could life exist on Mars without "life chemicals"?

This has led scientists to conclude that, at least at the sites of the Viking landers, life does not exist in the soil. Rather, the soil contains active chemicals, called super-oxides, which liberate gases when mixed with water (as hydrogen peroxide liberates oxygen when mixed with water). This would account for the positive results from the first 3 experiments and the negative result from the 4th.

The Martian Satellites; In 1877, the American astronomer Asaph Hall discovered the only two known natural satellites of Mars. The satellites were named Phobos and Deimos ("fear" and "panic"); appropriate names for the companions to the god of war (Figures 3.38 and 3.39).

Phobos, the innermost satellite, is almost 6,000 miles from the center of Mars and orbits in about 7 1/2 hours. Deimos is 14,000 miles from the center of Mars and has a period of just over 30 hours. Because of its rapid orbital motion, Phobos would appear, to a Martian observer, to rise in the west and set in the east.

The large crater on Phobos is named "Stickney", which was the maiden name of Hall's wife. The parallel grooves on Phobos may be related to the formation of this crater.

Photos of the satellites from the Viking orbiters show that each is elongated and cratered. Phobos is 17 miles long by about 12 miles wide and Deimos is about 10 miles long by about 6 miles wide. It is believed that both satellites were originally part of the asteroid belt and were captured by Mars early in its history.

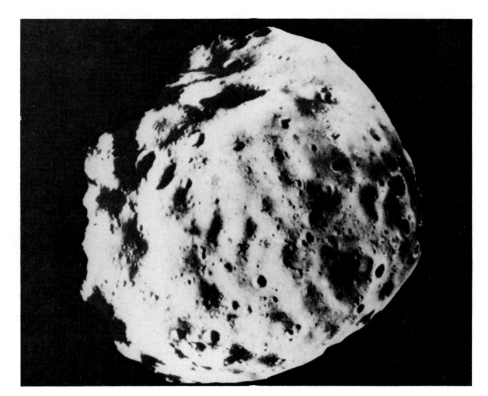

Figure 3.38: Phobos

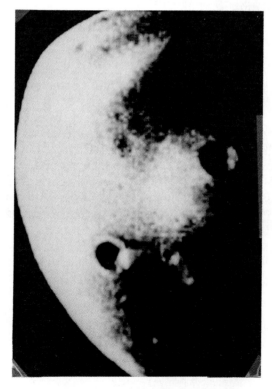

Figure 3.39: Deimos

JUPITER

Beyond Mars, we enter the realm of the gas/liquid, giant planets. The **Jovian planets** (they are named for Jupiter, the prototype) Jupiter, Saturn, Uranus and Neptune exist in stark contrast to the **Terrestrial (Earth-like) planets** Mercury, Venus, Earth, and Mars. While the Terrestrial planets are small rocky worlds with relatively high density, the Jovian planets are composed principally of hydrogen, helium, ammonia, methane and water. They are very large and have relatively low densities. Jupiter is their leader.

Jupiter is a rewarding object when viewed through a small telescope. Through such an instrument, one easily sees clouds in the Jovian atmosphere and the planet's four largest satellites. These are the same satellites discovered by Galileo in 1610; and are now known as the **Galilean satellites**.

As seen from the Earth, the clouds of Jupiter resolve into parallel bands of light and dark material. The white bands are called **zones** and the dark reddish bands are called **belts** (Figure 3.40).

Figure 3.40: Voyager Montage of the Jupiter System

The yellowish/white zones are thought to be composed primarily of ammonia ice crystals while the reddish belts seem to be made up of compounds of ammonia ("ammonium hydrosulfides"). The overall composition of the atmosphere of Jupiter, however, is 98 percent hydrogen and helium (78 percent hydrogen and 20 percent helium); the same as the composition of the atmosphere of the Sun.

General Properties of Jupiter and Its Orbit; Jupiter is the largest planet in the solar system with a diameter of 88,000 miles (11 Earth diameters). Its distance from the Sun is 480 million miles (5.2 au) and it has a sidereal period of revolution of just under 12 years. From the orbital motion of its satellites, the mass of the planet is calculated to be 318 Earth masses. This makes Jupiter more than twice as massive as all the other planets combined. Jupiter comprises 71 percent of the total planetary mass. From its mass and volume, the density of the planet can be calculated. This turns out to be about 1.3 grams/cc; about 30 percent more than the density of water. This indicates that, percentage-wise, Jupiter is composed mostly of the light elements, hydrogen and helium.

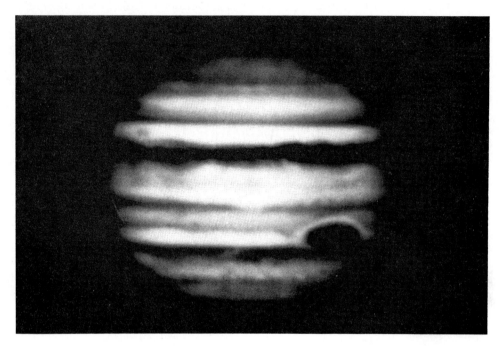

Figure 3.41: Jupiter as Seen From Earth

Jupiter is the most rapidly rotating planet; turning once every 9h 50m. This rapid spin is probably the reason why the clouds of Jupiter are drawn out into parallel bands (belts and zones). The planet is turning so rapidly that a point on the equator has a speed of about 25,000 miles per hour. The rapid equatorial speed makes the planet "oblate" (non-spherical); the polar diameter is 7 percent less than the equatorial diameter. An Earth-based photo of Jupiter is shown in Figure 3.41.

In the mid-1950's, it was discovered that intense radio radiation was coming from Jupiter. This radiation is emitted from charged particles trapped in Jupiter's magnetic field. The strength of the radiation indicates that Jupiter's magnetic field is intensely strong; more energetic than any field in the solar system except the Sun's.

Spaceprobes to Jupiter; In all, 4 spaceprobes from the Earth have been sent to Jupiter. In 1973 and 1974, Pioneer 10 and 11 visited the giant planet. In 1979, Voyager 1 and Voyager 2 passed Jupiter on their way to the outer solar system. These craft contained TV cameras as well as instruments to analyze planetary temperatures, atmospheric compositions and planetary magnetic fields (Figure 3.42).

The Voyager spacecraft revolutionized our ideas about Jupiter and the planets beyond Jupiter. The image resolution from its TV cameras was 100 times better than Earth-based photography. Most of the following information comes from Voyager 1 and 2.

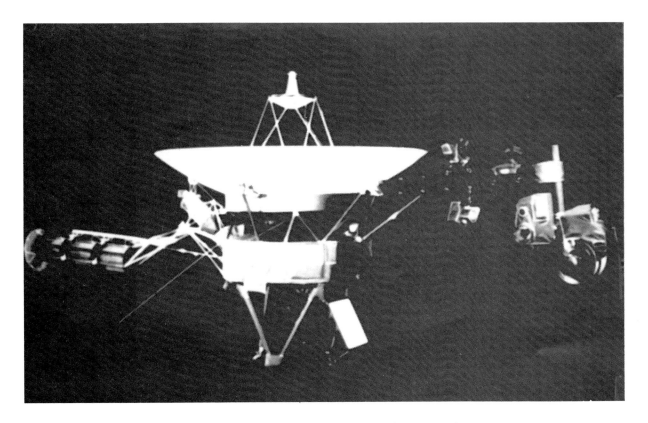

Figure 3.42: The Voyager Spacecraft

The Great Red Spot; The most prominent Jovian feature seen from the Earth is a pinkish-orange oval some 25,000 miles long by 10,000 miles wide. The **Great Red Spot** was discovered in 1665 by Cassini and has been seen almost continuously ever since (there are periods of time when the spot fades and is difficult to see). It lies about 5 miles above the tops of the surrounding clouds (Figure 3.43).

Voyager's cameras revealed that the red spot is a vast anticyclonic storm system with a rotation period of about six Earth days. While only the Great Red Spot can be seen from the Earth, Voyager revealed numerous, smaller red and white spots which are probably also storm systems in the Jovian atmosphere. These storms give scientists a chance to study the meteorology of a planet other than Earth; a second "data point" in our quest to understand planetary atmospheres (Figure 3.44).

The Atmosphere of Jupiter; As previously mentioned, the top of Jupiter's atmosphere consists of alternating light and dark zones and belts. The light zones are composed of regions of rising atmosphere, topped by clouds of ammonia ice crystals. The zones are about 10 to 15 miles higher than the dark belts which are regions of falling gas. The belts are also about 20 F° warmer than the zones.

These rising and falling "convection" currents on Jupiter are driven by the heat escaping from the hot interior of the planet. The reason that the currents are drawn out into parallel bands is that the planet has such a rapid rotation.

Figure 3.43: Jupiter from Voyager

Figure 3.44: Voyager Photo of the Great Red Spot

The temperature at the Jovian cloud tops is about -240° F and as one descends though these clouds, the temperature increases. The top cloud layer is composed of the white ammonia ice crystals. Below this level there is a reddish layer of ammonium hydrosulfide clouds. As we descend still further and the temperature rises, we should pass through water ice clouds, and below these, water vapor clouds like those on Earth. At this level, the temperature and pressure would be similar to that on Earth although the atmospheric composition would be vastly different (Figure 3.45). (Actually, at the point where the pressure is one atmosphere*, the temperature is approximately -200° F and where the temperature increases to 50° F, the pressure has become 10 atmospheres.)

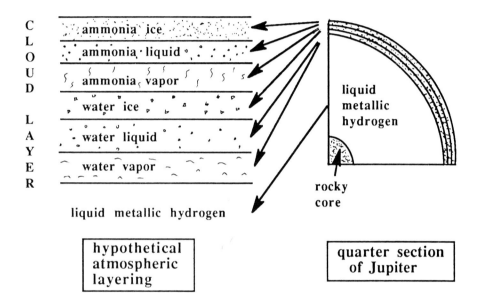

Figure 3.45: The Cloud Layers of Jupiter

As one continues downward, the temperature and pressure rise. On Earth, descending through the atmosphere, one eventually encounters a solid surface. This is not the case with Jupiter! After falling through 700 miles of atmosphere, one encounters the body of the planet; a hot dense ocean of liquid hydrogen.

The "Body" of Jupiter; The structure of the body of Jupiter may be best described as a ball of liquid hydrogen 80,000 miles in diameter.† The reason that the hydrogen is liquified is because of the tremendous pressures as one goes deeper into the planet. Computer models indicate that, at the bottom of the sea of hydrogen, the temperature is about 50,000°F, the pressure is 100 million Earth atmospheres and the density is over 3 times the density of iron!

It is further believed that at the center of the planet is a small core of heavy elements (iron, nickel, etc.), which makes up about one percent of Jupiter. This heavy material would naturally sink to the planetary center. The core is probably about twice the diameter of Earth.

Jupiter radiates into space more than twice the amount of energy it receives from the Sun. This energy is heat stored in the planetary interior and originated as energy released when Jupiter contracted from a gas cloud into a planet. For the last five billion years, Jupiter has been slowly radiating this energy into space (internal cooling).

*An "atmosphere" refers to the Earth's atmospheric pressure.
†Actually, the body of Jupiter is about 60 percent hydrogen and 40 percent helium plus silicates (by mass).

The Satellites of Jupiter; Jupiter has 16 satellites and a faint ring system. The four largest satellites are those discovered by Galileo in 1610 (the Galilean satellites). The inner two (**Io** and **Europa**) are about the size of the Moon and the outer two (**Ganymede** and **Callisto**) are about as large as Mercury (Figure 3.46).

Four very small satellites orbit Jupiter inside the orbit of Io. Beyond Callisto there are four small satellites which orbit in prograde orbits with a high inclination and four more small objects which have retrograde orbits. These outer eight satellites are probably captured from the asteroid belt.

Figure 3.46: The Galilean Satellites of Jupiter

Io: The innermost large satellite of Jupiter is Io. Its density is similar to the Moon in that it is composed of predominantly rocky material and little of the light material that characterizes Jupiter. Io is the most volcanically active body in the solar system. Eight volcanoes were erupting when Voyager 1 passed the satellite in 1979 and six were still active when Voyager 2 passed four months later (Figure 3.47).

The volcanoes of Io are ejecting sulfur and sulfur dioxide. This material covers the surface of Io with a blanket of sulfur and sulfur compounds; chemicals which, depending on temperature are black, white, red, orange and yellow. This gives Io its characteristic coloration (Figure 3.48).

The internal activity of Io is a result of tidal forces (see "Tides", Chapter 5) between Io and Jupiter. As Io swings about Jupiter, Jovian tides cause the satellite to be stretched into an elongated form. Since the orbit of Io is elliptical, the tidal forces vary as the Io-Jupiter distance changes. The changing tidal force causes a change in the size of the tidal bulge. This "pumping" of the body of the satellite causes the heating which results in the volcanic activity.

Europa: As with Io, Europa has a mainly rocky composition. Unlike Io, Europa is totally covered with frozen water. The lack of impact craters indicates that the surface "heals" itself; probably filling in cracks and holes with ice. The most prominent surface markings are cracks in the ice which have apparently been filled in (Figure 3.49).

Figure 3.47: Io and an Active Volcano

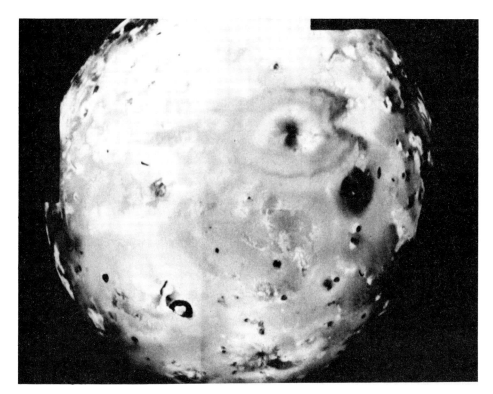

Figure 3.48: Surface of Io

Figure 3.49: Europa

It has been suggested that beneath its frozen surface (the ice is 80-100 miles thick), the ocean of Europa may be liquid water. This has not been verified.

Europa is the brightest of the jovian satellites; as bright as a sheet of white paper or perhaps as white as snow or frost.

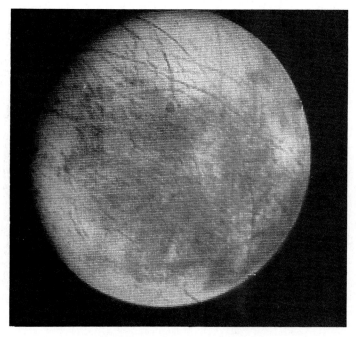

Figure 3.50: Europa

Ganymede: Ganymede is the largest satellite in the solar system (diameter = 3200 miles) and has a density slightly greater than that of Jupiter. This indicates that the satellite is made of an ice/rock combination. The satellite has thin water polar caps and a negligible atmosphere.

Figure 3.51: Ganymede

The surface of Ganymede is varied; about 1/3 being heavily cratered (an indication that the surface is ancient) while the rest is mountainous and covered with regions of frozen water. The younger terrain was formed when internal forces ruptured the surface, forcing up mountains and flooding regions of the surface with crustal water (Figure 3.52). These eruptions took place two to four billion years ago.

Displacement of sections of the crust are analogous to continental drift on the Earth. This displacement probably occurred about one billion years after the satellite formed; the same time the mountains formed (Figure 3.53).

Callisto: Callisto is just under 3000 miles in diameter and has a surface which is totally covered with impact craters. The low density of the satellite (less than 2 grams/cc) indicates that we are dealing with an icy, rather than rocky, world. Despite the icy nature of Callisto, it has been able to preserve its cratered surface (Figure 3.54).

Figure 3.52: Ganymede

Figure 3.53: Ganymede

Callisto has had little, if any, geologic activity as evidenced by the preservation of its ancient surface. The crater walls on Callisto are not as high as on the Moon. This may be due to the fact that its icy crust cannot support structures as high as can the solid crust of the Moon.

The surface seems to be covered with dark soil except where impacts (craters) have blown it away revealing an icy/bright sub-surface. The surface is probably 4 to 4 1/2 billion years old.

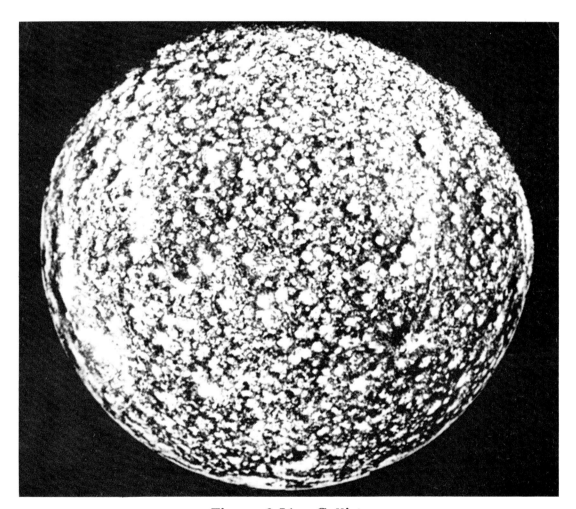

Figure 3.54: Callisto

The Other Satellites of Jupiter; The eight outer small dark satellites of Jupiter range from 5 to 100 miles in diameter. Four of them are about 8 million miles from Jupiter in prograde orbits with a 27° orbital inclination to the planet's equator. Four are about 15 million miles from Jupiter in retrograde orbits with a 52° inclination. Each group of 4 may have been a single asteroid which broke up during the time that it was being captured by Jupiter.

The four satellites inside of the Galilean satellites are of the order of 25 miles in diameter except for **Amalthea** which is about the size and shape of New Jersey (150 miles by 100 miles).

The Ring of Jupiter; While passing Jupiter, Voyager discovered a tenuous dust ring girdling the equator of the planet. Not as extensive as the rings of Saturn, Jupiter's ring appears to be made of particles eroded from the inner satellites of the planet. The ring represents an equilibrium between particles added to, and lost from, the region (Figure 3.55).

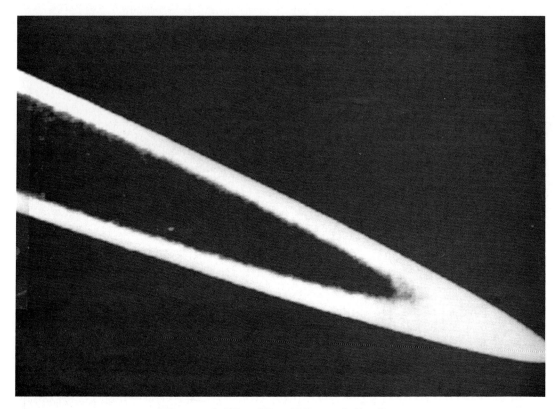

Figure 3.55: The Ring of Jupiter

SATURN

The most beautiful object which can be viewed through a small telescope is undoubtedly the planet Saturn. Its extensive ring system makes this planet a unique visual experience. While the rings of Saturn were observed by Galileo in 1610, the poor quality of his telescope made it difficult for him to interpret what he saw. In 1656, the astronomer Christian Huygens identified the "ring" around the planet. In 1675, Giovanni Cassini observed a "gap" in the ring (now called **Cassini's division**) which showed that there were are least two rings (Figure 3.56).

Figure 3.56: Saturn from Earth

Saturn is the second largest planet in the solar system; its diameter is nine times that of the Earth and its mass is 95 Earth masses. It is, in most ways, a little brother to the planet Jupiter and many of its properties are the same as Jupiter's.

The Revolution and Rotation of Saturn; Saturn is 9.5 au from the Sun and has a period of revolution of 30 Earth years. The planet rotates in just under 11 hours; a little more slowly than Jupiter. Like Jupiter, the exact rotation period is a function of latitude on the planet; the equatorial period being shorter than at places either north or south (the equatorial period is 10h 14m while the polar regions rotate in 10h 40m).

The Mass and Density of Saturn; While Saturn has about half the volume of Jupiter, it has less than 30 percent as much mass. This means that the density of the planet is very low; 0.7 grams/cc; the lowest density of the Jovian planets. In fact, since the density of Saturn is less than 1 gram/cc, the planet would float in water!

Figure 3.57: Saturn with Several of its Satellites (Voyager 1 Photo)

Space Probes to Saturn; While Pioneer 11 reached Saturn in 1979, the information returned by Voyager 1 in 1980 and Voyager 2 in 1981 was so superior that Pioneer is largely ignored. Pioneer is best remembered as being the first Earth probe to leave the Solar System. Figure 3.58 shows the plaque carried aboard Pioneer; a message from Earth to the Universe.

Figure 3.58: The Pioneer Plaque: On its Way to the Stars

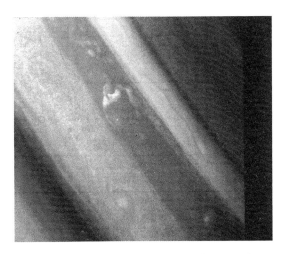

Figure 3.59: Saturn as seen from Voyager

Figure 3.60: Voyager Photo of Saturn

Images of Saturn returned from the Voyager spacecrafts show a cloud structure with much less detail and contrast compared to Jupiter. This is true since at the colder temperatures of Saturn (-280° F) less organic compounds can form. Even so, zones and belts, as well as storm patterns could be discerned. 1000 mile/hour winds (4 times higher than the winds on Jupiter) were measured in the Saturnian atmosphere. As with Jupiter, Saturn gives off radio waves and about two to three times as much energy as it receives from the Sun. The radio signals are an artifact of Saturn's intense magnetic field and the excess energy indicates that the planet, like Jupiter, has a hot interior.

Saturn has a liquid hydrogen interior extending to the surface of a rocky core made of heavy material. The core is thought to be about twice the diameter of the Earth and contains about 11 Earth masses.

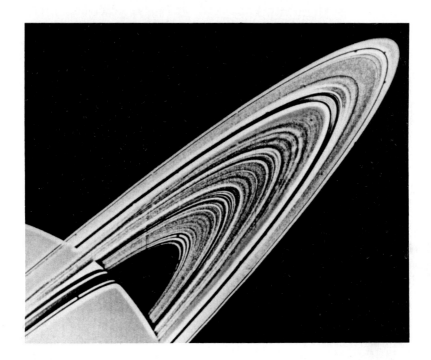

Figure 3.61: Voyager Image of the "Ringlets" of Saturn

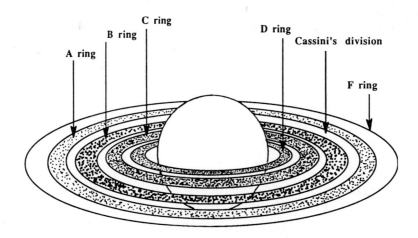

Figure 3.62: Designation of Saturn's Rings

The Rings of Saturn; From the Earth, it appears as though Saturn has three rings. The Voyager spacecraft indicated that this was far from the true picture. In actuality, there are hundreds or thousands of **ringlets**, each composed of a myriad of small pieces of water ice in orbit about Saturn. The "ice balls" which make up the rings vary from golf ball size to house size (Figure 3.61). Saturn's ring system is about 171,000 miles in diameter and less than 100 miles thick. The ring system is so thin that once every 15 years, when the plane of the rings is edge on to Earth, the rings disappear for a few days.

The largest "gap" in the rings of Saturn is called **"Cassini's division"** (Figure 3.62). The outside ring (outside Cassini's division) is known as the **"A"** ring. The ring just inside the division is known as the **"B"** (bright) ring. Fainter rings are called the **"C"**, **"D"** and **"F"** rings (Figure 3.62).

It is believed that the ringlets may be formed and maintained by the action of Saturn's satellites, which either gravitationally or physically "shepherd" the ring particles into ringlets.

There is much that we do not understand about the rings of Saturn. Why are there individual ringlets? Why are some ringlets not circular? The rings show dark shadows called "spokes" which are thought to be caused by dust particles suspended above the ring plane (Figure 3.63). What is the source of this dust?

Figure 3.63: "Radial Spokes" in Saturn's Rings

The origin of the rings of Saturn is also not certain. Since the rings lie within **Roche's limit** of Saturn (that distance within which tides from a planet would tear a satellite apart), the ring particles either represent a satellite that wandered in too close and was torn apart by Saturn's gravity, or material which was prevented from condensing by tidal forces.

The Satellites of Saturn; While only about 10 satellites of Saturn were known before the Pioneer and Voyager probes, about 2 dozen have now been observed. The largest of these, **Titan**, is larger than Mercury and has an extensive atmosphere.

Figure 3.64: Collage of the Major Satellites of Saturn

We will discuss the major satellites in order of their distance from Saturn. Just outside the rings are five small satellites between 12 miles and 150 miles in diameter. These satellites were discovered by Voyager.

MAJOR SATELLITES:

Mimas: This satellite is about 250 miles in diameter and possesses one large impact crater. The impact which formed this crater was almost violent enough to fragment the satellite. The density of the satellite is about 1.2 grams/cc and it is, therefore, composed mostly of water ice (Figure 3.65).

Figure 3.65: Mimas

Figure 3.66: Enceladus

Enceladus: This satellite is about 310 miles in diameter, has a density of 1.1 grams/cc (it is almost all water ice), and is covered with straight grooves similar to those on Ganymede. It has the brightest surface in the solar system. This surface has been modified by eruptions of H_2O from a hot interior. The heating of the interior may be due to tidal interactions with other satellites (similar to the situation with Io). (Figure 3.66)

Tethys, Rhea, Dione: These satellites range in diameter from 600 to 1000 miles and have bright icy surfaces. Tethys is 100 percent ice. Dione and Rhea have rock mixed with ice and heavily cratered surfaces. This indicates that the surfaces are old and that there is little or no internal activity (Figure 3.67).

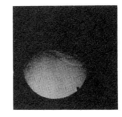

Figure 3.67: Tethys, Rhea, Dione

Two small satellites (12 to 13 miles in diameter) are co-orbital (share the same orbit) with Tethys and are found 60° ahead and 60° behind the ice satellite. One small satellite is co-orbital with Dione.

Titan: The largest satellite of Saturn is Titan. It has a diameter of 3,200 miles (it is larger than Mercury) and is the second largest satellite in the solar system. Titan has an extensive atmosphere which consists of 90 percent nitrogen and 10 percent methane gas plus other trace gases. The atmospheric surface pressure is 60 percent higher than Earth's surface atmospheric pressure. The surface temperature of -300°F is cold enough for liquid and solid methane to exist. It is theorized that an ocean of liquid methane up to 1/2 mile deep may cover the satellite and that methane rain and snow might fall from time to time. In these qualities, methane on Titan takes the place of water on Earth.

Because the clouds of Titan are opaque, we have not been able to see down to the surface. Even though nitrogen is generally transparent, solar photochemical action produces a form of smog which hinders visibility. This smog is made of complex organic molecules that may actually rain down on the surface of the satellite creating an organic "soup" similar to that in which life formed on the early Earth (Figure 3.68).

Hyperion: This is an oblong satellite (210 miles x 120 miles) which has a rotation rate which is irregular. This is known as "chaotic rotation" (Figure 3.69).

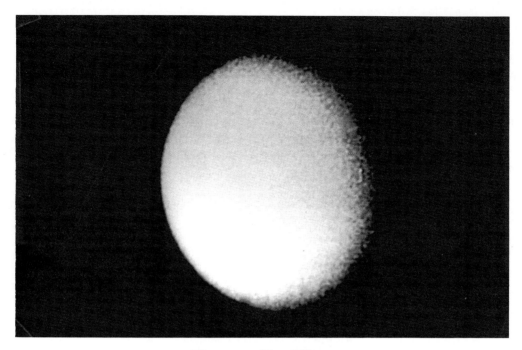

Figure 3.68: Titan

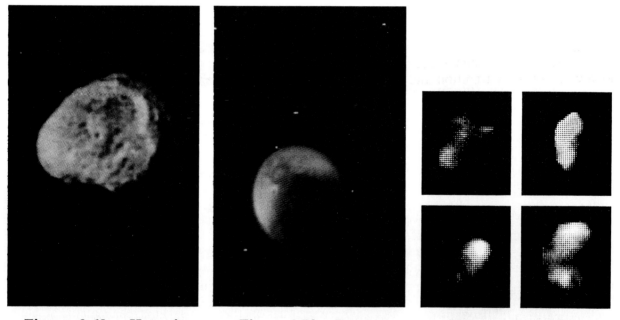

Figure 3.69: Hyperion   **Figure 3.70: Iapetus**   **Figure 3.71: Phoebe**

Iapetus: This large satellite (900 miles in diameter) has a "leading" edge* covered with dark material and a "trailing" edge covered with bright white ice.

Phoebe: Phoebe is 140 miles in diameter. This is a <u>very</u> dark satellite and has a retrograde orbit. It is possible that black dust knocked off Phoebe by meteorite impact would spiral in towards Saturn and be swept up by the leading edge of Iapetus.

*The "leading edge" of an orbiting body is the hemisphere of the object which is moving forward through space.

URANUS

While much larger than the Terrestrial planets, both Uranus and Neptune are less grand than the Jovian planets inside their orbits. Both are about 30,000 miles (about four Earths) in diameter and about 15 times as massive as the Earth. They are both made principally of hydrogen and helium with a small proportion of heavier elements thrown in.

The Discovery of Uranus; Uranus was discovered in 1781 by the English astronomer **William Herschel**. When he first observed it, its disk-like appearance caused him to believe that it was a comet. When its orbit was computed, however, it was found to be almost circular; something more planet-like than comet-like. Since no new planet had ever been discovered, the realization that this object was a planet came as a great surprise.

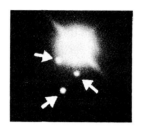

Figure 3.72: Earth-based Photo of Uranus and its Major Satellites

Herschel proposed to name the new planet "Georgium Sidus" (George's star) in honor of King George III. Some wished to name it Herschel, in honor of its discoverer. It was finally decided, however, to follow tradition and name it after one of the gods of Roman mythology. Uranus, grandfather of Jupiter, was ultimately chosen.

Uranus is at an average distance of 19 au from the Sun and completes one orbit in 84 years. The best Earth-based observations show no detail on the planet. Spectroscopic evidence suggests clouds of methane in the upper atmosphere. The combination of the methane with hydrogen gives Uranus a bluish/green tint which is visible from Earth.

The period of rotation of Uranus is just over 17 hours. A unique property of Uranus is that the inclination of its rotation axis is 82° and its rotation is retrograde. This situation (Figure 3.73) causes one Uranian hemisphere to face the Sun for 42 years and the opposite hemisphere to face the Sun for the next 42 years. The phenomenon is an extreme case of the Earth's polar "midnight sun".

The "Body" of Uranus; While both Uranus and Neptune are Jovian planets, they contain higher percentages of the heavier elements than do Jupiter and Saturn. This is known because their densities (1.3 g/cc for Uranus and 1.7 g/cc for Neptune) are greater than those of Jupiter and Saturn. Despite this, these two planets are primarily hydrogen and helium. Spectroscopic evidence points to the existence of methane clouds in the upper atmosphere (at the temperature of the upper atmosphere, about -360° F, methane would form clouds of ice crystals). These methane ice clouds are efficient at absorbing red and orange light and give Uranus and Neptune their characteristic blue-green colors. According to some theories, the "bodies" of both Uranus and Neptune are composed of a dense, slushy, hydrogen/helium/water mixture all the way down to an Earth-sized iron/silicon core similar to those of Jupiter and Saturn.

The upper atmosphere of Uranus is marked by extremely faint east-west cloud bands, similar to the clouds in the atmospheres of Jupiter and Saturn. These clouds are caused by the planet's upper atmospheric "jet stream" which blows at speeds up to 350 miles/hour (compared to about 100 miles/hour for Earth). (Figure 3.73)

Figure 3.73: Uranus from Voyager 2

The 17 hour rotation period of Uranus was measured by observing variations in radio emissions as solar wind particles interact with the planet's magnetic field. One surprising result of these experiments was that Uranus' magnetic field was found to be tipped to the rotation axis by about 60°. This is a much larger value than usual (the value for Earth is less than 12°), and may indicate that Uranus' magnetic field is undergoing a reversal or that Uranus suffered a catastrophic collision some time in the past (Figure 3.74). The magnetic field of Uranus is about 50 times as strong as the Earth's field.

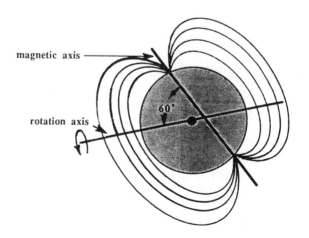

Figure 3.74: The Magnetic Field of Uranus

Voyager 2 has discovered that the center of the magnetic field is <u>not</u> at the center of the planet but displaced by about 5,000 miles from the center, in the mantle of the planet.

The Rings of Uranus; In 1977, Uranus was to occult (eclipse) a distant star. Astronomers saw this as an opportunity to get an accurate measure of the diameter of the planet. This would be done by observing the length of time that the star was blocked by the disk of the planet. As the event was only visible from the Indian Ocean, measurements were made from the **Kuiper Airborne Observatory**, an instrumented jet aircraft.

An unexpected observation was made about 30 minutes before and after the expected occultation by the planet, a series of small "dips" in brightness was observed (Figure 3.75). These dips indicated that Uranus was surrounded by at least 5 narrow rings not directly visible from Earth.

Later studies (Earth based and from Voyager 2) have revealed that Uranus has a total of 11 rings. Unlike the rings of Saturn, which are broad and bright, the rings of Uranus are narrow (generally less than six miles wide) and quite dark. The particles which make up the rings are about three feet across and have the brightness of a lump of coal! The rings are all within Roche's limit of the planet's center. The narrowness of the rings may be caused by the action of shepherd satellites, but only one or two have been discovered so far (Figure 3.76).

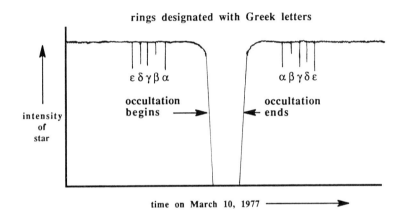

Figure 3.75: Discovery of the Uranian Rings

Figure 3.76: The Rings of Uranus

It has been suggested that the rings of Uranus are a transitory phenomenon, eroding as we watch them, and that they might be gone in a relatively (astronomically speaking) short time.

The Satellites of Uranus; Voyager 2 passed through the Uranian system on January 24, 1986. The craft passed within 50,000 miles of the planetary surface and made discoveries concerning not only Uranus, but the satellite system as well. From the Earth, five satellites of Uranus can be seen. The two largest, **Titania** and **Oberon**, were discovered by Herschel in 1789. Two others **Ariel** and **Umbriel** were discovered in 1851 and **Miranda** was discovered by Kuiper in 1948. The five satellites were named after sprites and spirits from English literature. Titania and Oberon are about 1,000 miles across, while Ariel and Umbriel are about 750 miles in diameter. Miranda is about 300 miles in diameter (Figure 3.77).

During its passage, Voyager 2 discovered 10 small satellites. All, except for **Puck**, are less than 60 miles in diameter.

Figure 3.77: Uranus and its Major Satellites: Titania, Oberon, Umbriel, Ariel, Miranda

All of the Uranian satellites are rather dark and may be coated with the same soot-like material which coats Iapetus and the ring particles of Uranus. The densities of the satellites are all about 1.5 g/cc and are, therefore, probably a mixture of water ice and rock.

Miranda; The most interesting of the Uranian satellites is Miranda (Figure 3.78). It appears to be constructed of several large block-like sections. Some of these blocks are composed of ice, while others are composed of rock. The surface is a dramatic array of jumbled topography.

It is theorized that Miranda originally consisted of a rocky core and an icy mantle. Then, some time in its history, a collision shattered the satellite into blocks of rock and ice. The impact, however, was not large enough to separate the blocks permanently and gravity subsequently brought the blocks together to form the jumbled conglomerate we now see.

An alternate explanation for the jumbled surface of Miranda is that the planet, during its formative period, froze before it could completely gravitationally differentiate itself into a rocky core and icy mantle. The most interesting feature on the surface of the satellite is a cliff 16 miles high! A person, jumping off this cliff, would take 9 minutes to hit bottom!

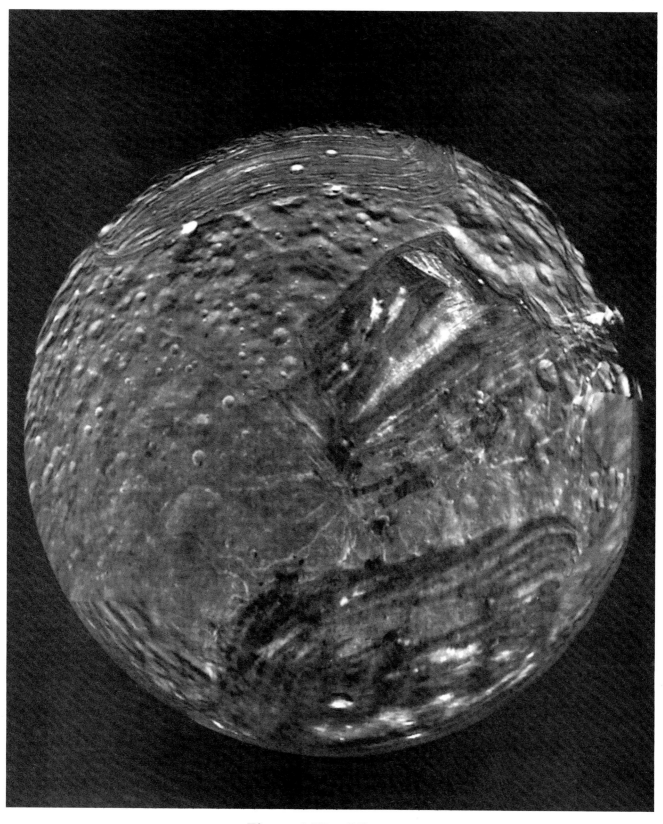

Figure 3.78: Miranda

NEPTUNE

After the discovery of Uranus in 1781, astronomers computed its theoretical orbit from Newton's laws of motion and Newton's law of gravity. Scientists of that time saw Uranus as an opportunity to check the validity of the rather new Newtonian physics. By the 1830's, however, it was clear that Uranus was not moving as predicted. For several years it was in advance of its theoretical position and then, later, behind its theoretical position. Some thought that this signaled an error in Newton's laws.

In 1843, **John Couch Adams**, a 24 year old Cambridge University student, began to mathematically explore the possibility that the deviant behavior of Uranus might be due to the gravitational effect of an, as yet undiscovered, eighth planet. From the irregularities in the motion of Uranus, Adams was able to use Newtonian theory to predict the exact position of the hypothetical planet (The prediction was that the planet was in the constellation of Aquarius.). In 1845 Adams submitted his prediction to Sir George Airy, the Royal Astronomer of England, but Airy had little faith in such a young and unknown mathematician and he summarily dismissed the prediction.

In 1846, a French astronomer named **Urbain Jean Joseph Leverrier** performed the same calculation as did Adams and he arrived at the same result. When Airy, in England, read Leverrier's paper on the subject, he realized that he might have blundered by ignoring Adams' prediction. He contacted James Challis, the director of the Cambridge University Observatory, and requested that he begin the search. Because of both a lack of aggressiveness in the search and a lack of up to date star charts of the predicted region in Aquarius, the search was never completed. Meanwhile, Leverrier had contacted Johann Gottfried Galle at the Berlin Observatory. The observatory happened to possess good star charts of the region of the sky in question so that the presence of an "interloper" would be immediately noticed. Galle received Leverrier's communication on September 23, 1846 and that very night, within 20 minutes of beginning the search, the planet was discovered.

Leverrier proposed that the planet be named after Neptune, the Roman God of the Sea. After years of debate, it has been agreed that credit for the discovery of the planet be given jointly to Adams and Leverrier.

As it turns out, Galle was probably not the first person to observe Neptune. In 1612 and 1613, Galileo was making observations of Jupiter and its four major satellites. Galileo made sketches of the nightly motion of these satellites and the background stars. One night Galileo noted (in writing) that one of the background "stars" had appeared to move. Calculations indicate that Neptune, in the winter of 1612-13, was in the position of that "star". Neptune, therefore, may have been observed 234 years before it was "discovered" by Adams and Leverrier!

The Orbit of Neptune: Neptune orbits at a mean distance of just over 30 au from the Sun. At this distance, its period of revolution is 165 years. While it is considered the "8th" planet, the extreme eccentricity of Pluto's orbit sometimes brings the "9th" planet in closer to the Sun than Neptune. This situation is occuring now and will last until 1999. More will be said about this in the section on Pluto.

The "Body" of Neptune: The structure of Neptune is apparently quite similar to Uranus. It has a diameter of approximately 30,000 miles and a mass of just over 17 Earth masses. Its surface gravity is just over 1g (g = Earth gravity) and has a density of about 1.6 grams/cc. The temperature at the tops of the Neptunian clouds is about -360°F.

The planet almost certainly has a rocky core similar in size to the Earth. This core is overlain by a dense, slushy atmosphere of superdense hydrogen and helium with dense "clouds" of water into which much of the planet's ammonia might be dissolved. The outer cloud layer of the planet is a mixture of low density, low pressure, low temperature hydrogen and helium. As with Uranus, the highest clouds are composed of methane ice crystals whose ability to absorb red light causes the planet to exhibit a blue color.

In August, 1989, Voyager arrived at Neptune. Photos of the planet show a blue world marked by high, white, methane clouds and a large dark area which is known as the **Great Dark Spot** (Figure 3.79).

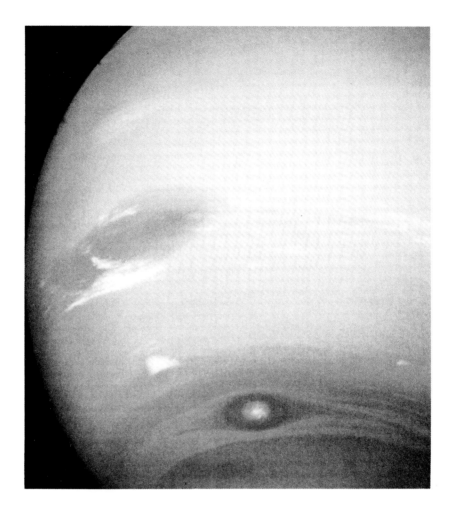

Figure 3.79: Neptune from Voyager 2

The Great Dark Spot is an oval structure with a long axis about 8,000 miles across (about the diameter of Earth). The dark spot is a storm system similar to Jupiter's red spot and is, proportionaly, the same size as the red spot (and at the same latitude). As with the red spot, the dark spot is anti-cyclonic in nature and is, therefore, a high pressure system. The white clouds above the dark spot are similar to stationary lenticular clouds on Earth. The winds above the dark spot are moving at about 1,500 miles/hour! The white methane clouds of Neptune are about 30 to 40 miles above the blue cloud deck.

As with Jupiter, Neptune radiates more energy than it receives from the Sun. The heat is thought to be a remnant of condensation of the planet 4.6 billion years ago.

A study of radio emissions from the planet revealed a 16 hour periodicity. This is almost certainly the rotation period of Neptune. A unique feature of Neptune is that while the planet takes 16 hours to rotate, the atmosphere takes 18 hours to rotate. This means that relative to the planet, the atmosphere moves retrograde!

Reminiscent of Uranus, the magnetic axis of the planet is steeply inclined to the rotational axis of the planet. In this case, the angle is 47°. Also, as with Uranus, the magnetic axis does not pass through the planet's center.

The Rings of Neptune: Neptune has a series of rings; four in all. Two of the rings are narrow (less than 30 miles wide) and two are wide (1,000 miles wide and 3,000 miles wide). The two wide rings are thin, low density sheets of particles.

Neptune's ring particles are quite dark and are probably made of water ice covered with some sort of carbon soot. It has been found that the particles in the narrow rings have tended to clump into concentrations which, from a distance, give the rings the appearance of **arcs** rather than complete circles (Figure 3.80).

The Satellites of Neptune: Before the flight of Voyager 2, two satellites of Neptune were known. Voyager discovered six more whose diameters are all less than 250 miles. The largest satellite of Neptune, **Triton** (named for the son of Poseidon, Greek God of the Sea), was discovered in the same year as was Neptune (1846) and is one of the most interesting satellites in the solar system. It was immediately ascertained that the orbit of Triton was retrograde (i.e. as viewed from the north, the satellite orbited in a clockwise fashion) with a period of 6 days. There is no other major satellite in our solar system with a retrograde orbit. This probably means that Triton was captured by Neptune rather than having formed by accretion (see "The Origin of the Moon" in Chapter 5).

Furthermore, the tidal interaction between Triton and Neptune is causing the satellite to gradually spiral towards the planet (see "Tides" in Chapter 5). The mechanism which causes this is as follows. The tidal action on Triton from Neptune raises a tidal bulge on the planet. Because the satellite is retrograde, however, the bulge "lags" behind the satellite and the gravitational force at the bulge on the satellite slows Triton's forward motion, causing it to slowly spiral towards Neptune. As we will see in Chapter 5, the Earth and the Moon are in the same sort of tidal relationship but since the Moon is in a prograde orbit, the effect is to speed up the Moon and cause it to spiral away from the Earth.

Figure 3.80: Ring "Arcs" of Neptune

Calculations indicate that within 100 million to 200 million years, Triton will close within 42,000 miles of Neptune* (it is presently 220,000 miles from the center of Neptune) at which point tidal forces from Neptune will tear the satellite apart and create a majestic ring system orbiting the eighth planet.

*This is Roche's limit for Neptune.

Triton is approximately 1,700 miles in diameter and is, therefore, slightly larger than Pluto. It orbits Neptune at an angle of about 20° to the Neptunian equator. This is similar to the angle at which the Moon orbits the Earth's equator and is an oddity since major satellites generally orbit their planet's equators. The probable "capture" origin of Triton (and, perhaps, the Moon as well) explains this situation.

Triton was photographed by Voyager 2 during its 1989 passage of the Neptunian system. Figure 3.81 shows a Voyager 2 view of Triton.

Figure 3.81: Triton

The polar region of Triton seems to be covered with a layer of "reddish" frost which has tentatively been identified as nitrogen frost (the temperature on Triton is almost -400°F!). Furthermore, the south polar region shows a scarcity of craters which implies that at least part of the cratered surface has been covered by an icy "lava". This occurred at some point since the bombardment that took place in the solar system during the time of its original condensation from the pre-solar nebula.

A logical explanation for the "lava" would be a heating of the interior of Triton caused by the tidal interaction between Triton and Neptune. This is the same tidal action which is causing the orbit of Triton to decay. Voyager 2's cameras also photographed what appears to be two **geysers** erupting on Triton. It is believed that these geysers are the result of nitrogen gas escaping through cracks in the satellite's surface.

Triton also exhibits large, flat areas similar to "frozen lakes". These areas may be the calderas of extinct volcanoes filled with ice "lava".

The other satellite of Neptune known before the flight of Voyager 2 is **Nereid**. Nereid has the most eccentric orbit of any satellite in the solar system and varies its distance from Neptune from six million miles to 900,000 miles. It takes one Earth year to orbit Neptune. It is only about 250 miles in diameter.

PLUTO

Pluto, the farthest known planet, is extreme in many ways. It has the most eccentric orbit of any planet, ranging from 30 au to 50 au from the Sun (the average is close to 40 au). Its orbit is so eccentric that it spends 20 years of each orbital period (its period of revolution is 248 years) inside the orbit of Neptune (Figure 3.82). Pluto crossed inside Neptune's orbit in 1979 and will remain there until 1999. Its orbit is the most inclined to the plane of the solar system (17°) and it is, by far, the smallest major planet (diameter = 1400 miles).

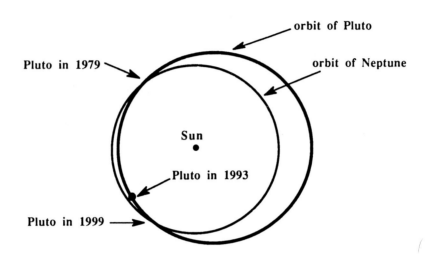

Figure 3.82: The Orbit of Pluto

Discovery of Pluto; After the discovery of Neptune, it was hoped that further planets could be discovered in the same manner; by their gravitational effects on inner planets. Taking up the quest, **Percival Lowell** decided to hunt for the ninth planet by looking for peculiar motions in the orbits of Uranus and Neptune. Since Neptune, by 1900, had not been observed for a long enough period for an accurate orbit to be computed, Lowell concentrated on perceived perturbations of Uranus' orbit.* Early in the 20th century, Lowell predicted, from his observations of Uranus, that the ninth planet had a mass of 6.6 Earth masses and was located in the direction of the constellation Gemini. Lowell searched for the suspected planet from his Flagstaff observatory from 1906 until his death in 1916.

Subsequently, the search continued under the general direction of Lowell's brother. In 1929 a new wide angle camera was installed to help in the search and a young astronomer, **Clyde W. Tombaugh**, was hired to assist in the screening of photographic plates taken of the suspected region.

On February 18, 1930, Tombaugh isolated, on two of the photographic plates, an object which was moving at the predicted rate (Figure 3.83). When an orbit was calculated, it was found to be orbiting about 40 au from the Sun. On March 13, 1930, the discovery of the ninth planet was announced to the world. The planet was named Pluto after the god of the underworld and also because the first two letters in Pluto are Percival Lowell's initials.

Pluto was much fainter than had been predicted and, therefore, probably smaller. Could it be that a larger "planet x" still existed at the fringe of the solar system. For the next 13 years Tombaugh continued the search but nothing was found. If there is another planet beyond the orbit of Neptune, it must be either very small or very far from the Sun.

*It was later shown that these perturbations had no statistical significance.

The distance of Pluto from the Sun is so great that even through the best Earth-based telescopes, the planet looks no larger than a pinpoint. It was, therefore, difficult to gather information about our distant neighbor. About the only thing which could be said with some certainty was that Pluto rotated once every 6.4 days. This was known since the planet dimmed and brightened with this period. Obviously, Pluto has bright and dark spots which are alternately brought into and out of view as the planet rotates.

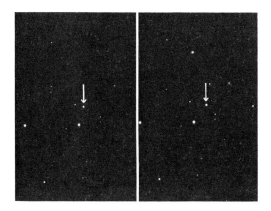

Figure 3.83: The Discovery of Pluto

The Discovery of Charon; Much of the mystery surrounding Pluto disappeared in 1978 when **James W. Christy**, of the U.S. Naval Observatory (in Flagstaff!), discovered a satellite orbiting Pluto. The satellite was named **Charon** (Kar'on) after the mythological boatsman who ferried the dead across the River Styx to Hades; Pluto's realm. The satellite was discovered when Christy, attempting to make precise measurements of the position of Pluto, dramatically enlarged a photo of Pluto (Figure 3.84).

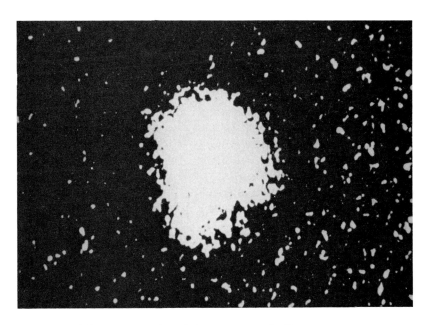

Figure 3.84: Discovery of Charon

It was noticed that the enlarged image of Pluto had a "bump" while other star images on the plate did not. Christy enlarged other photos of Pluto and found that the bump moved around from side to side, sometimes disappearing completely. The only explanation for this phenomenon was a close satellite orbiting Pluto. The reason that the two bodies could not be seen as separate objects was that the Earth's atmosphere "smeared" each image into overlapping disks. To see Pluto and Charon separately, photographs had to be taken from space. This was done in 1990-91 by the Hubble Space Telescope (Figure 3.85). The discovery of Charon has led to a revolution in our understanding of Pluto.

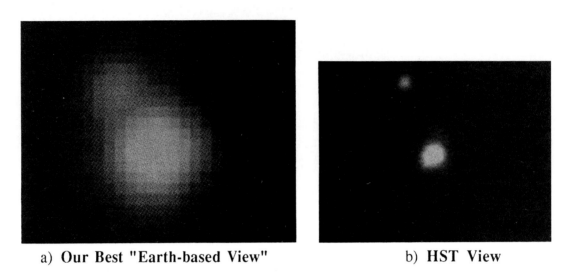

a) **Our Best "Earth-based View"**   b) **HST View**

Figure 3.85: Pluto and Charon as seen from Earth and by the Hubble Space Telescope

The Mass of Pluto; Observations of Charon have shown that it has a diameter of about 700 miles*, a distance of 12,200 miles from Pluto's center and a revolution period of 6.4 days (Pluto and Charon are in synchronous rotation).

The distance from Pluto to Charon and the revolution period can be used to calculate the mass of the Pluto-Charon system using Kepler's 3rd Law (Learning Activity 10, Appendix IV).

The result is that Pluto and Charon together comprise approximately 1/400th the mass of the Earth. But remember that the prediction by Lowell was based on a planetary mass of almost seven Earth masses. There is no chance that Pluto and Charon could have had a gravitational effect on Uranus or Neptune. Their mass sum is just too low! We must, therefore, reach the conclusion that the discovery of Pluto, although promoted by Lowell's enthusiasm, was based on pure chance!

The Surface of Pluto; The axis of rotation of Pluto is inclined to the axis of the solar system by 122°. Since Charon orbits the equator of Pluto, Earth-bound observers will see eclipses of Charon and Pluto during periods separated by 124 years (1/2 a Plutonian year). During these "eclipse seasons", eclipses occur every 3.2 days (Figure 3.86).

One of these "eclipse seasons" started in 1985 and lasted until 1990. During this period, as Charon alternately passed in front of and behind Pluto, scientists were able to make measurements of surface features of both bodies. It appears as though Pluto is brighter near the poles than at the equator. Pluto generally appeared brighter to us during the 50's because at that time we were

---

*Charon is the largest satellite in the solar system compared to the size of the planet about which it orbits. The barycenter of the Pluto-Charon system lies 750 miles above Pluto's surface. This has led some to call the pair a double planet.

observing the south polar region. At this time, we are seeing more of the equatorial region which consists of light and dark patches. This lowers the observed average brightness of the planet (Figure 3.87).

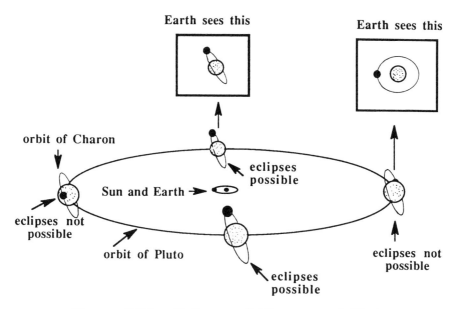

Figure 3.86: Eclipses of Charon and Pluto

In general, Pluto is brighter than Charon, reflecting some 50 percent of the incident sunlight back into space, compared to 40 percent for Charon. Spectroscopic studies during eclipses reveal that the surface of Pluto is covered by methane frost or ice while the surface of Charon is covered by a darker layer of water ice.

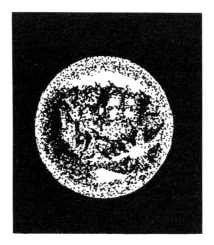

Figure 3.87: Artist's Conception of Pluto

In general, Pluto is seven times more reflective than our Moon (whose reflectivity is about 7 percent). This is due to the "white" nature of the frozen methane surface. But, over time, solar ultraviolet radiation and cosmic rays should cause the surface to darken. Why is it still so bright? One theory is that every time the planet reaches perihelion*, the increased solar radiation causes some of the surface ice to vaporize into the atmosphere (this happened last in 1989). When the planet draws farther from the Sun, towards aphelion, the lowering temperatures cause the methane to "snow out" onto the surface. This causes a re-whitening of the planet. The temperature at the surface of Pluto has been measured to be about -355°F.

The Atmosphere of Pluto; In 1988, Pluto passed between the Earth and a star. Rather than "winking out" instantaneously, the star light dimmed and <u>then</u> was quickly extinguished. This event allowed scientists to determine that Pluto has an upper clear atmosphere overlying a more opaque lower level. The gases are probably methane and nitrogen and the surface pressure is of the order of .00001 atmospheres.†

The "Body" of Pluto; Knowing the mass and volume of a planet allows one to compute its density. The density of Pluto is just over 2 grams/cc. This is a relatively high value for a body found so far from the Sun. Could Pluto have been formed in the inner solar system and then have wandered to its present position? Its density indicates a large rocky core with a water mantle covered by a methane crust. It is also interesting to note that Pluto is very similar to Triton, the largest satellite of Neptune. They have about the same diameter and density. Both appear to have formed as independent bodies and were later captured into their present orbits; Triton is a retrograde satellite and Pluto is a planet with a maverick orbit. Both Pluto and Triton were subsequently internally heated; Triton from its tidal interaction with Neptune, and Pluto by its gravitational interaction with Charon. Because both Triton and Pluto were heated, their rocky material could sink to the centers forming relatively dense cores. Some have suggested that Pluto might be an escaped satellite of the Neptunian system or rather might, one day in the future, collide with Neptune. The later event is unlikely as Pluto's orbit is in a 3:2 resonance with Neptune (three Neptunian years = two Plutonian years), and the two planets never get closer to one another than 1.7 billion miles.

---

*Perihelion refers to a planet's closest approach to the Sun.
†One "atmosphere" is the standard air pressure at the Earth's surface.

CHAPTER 4

THE MINOR MEMBERS OF THE SOLAR SYSTEM

The family of the Sun consists of more than just the major planets and satellites discussed in Chapter 3. Orbiting the Sun are hundreds of thousands of smaller, low mass bodies called minor planets, meteoroids and comets. While comprising a very small percentage of the total solar system mass, these objects play an important role in our understanding of the early history of our planetary system and of the forces which gave it its general properties. It has become clear that many of the small bodies wandering the depths of interplanetary space are made of the original material from which the solar system was created. Since some of these objects have changed little, if at all, during the last four and a half billion years, studying them could give us valuable information concerning our origin.

Also, the minor members of the solar system provide people on Earth with some of our most beautiful and exciting celestial displays; the flash of a meteor or bolide or the grace of a comet tail for example.

In this chapter we will explore the properties of the minor planets, comets and meteoroids and try to understand the events which lead to meteors, meteor showers, and cosmic collisions.

The Minor Planets; In the space between the orbits of Mars and Jupiter lies the **asteroid belt** or zone of **minor planets**. These objects range in size from 600 miles on down to a few hundred yards, and appear to be composed of rocky material, nickel-iron metal or a combination of rock and metal (Figure 4.1).

The discovery of the first and largest minor planet (named **Ceres** for the Goddess of Agriculture) by Giuseppí Piazzi on the first night of the 19th Century (January 1, 1801) came as no great surprise for astronomers who had been searching for several decades for the "missing planet" between Mars and Jupiter.

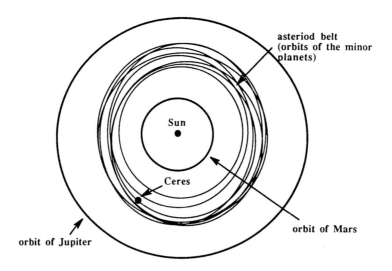

Figure 4.1: Distribution of Minor Planets

The reason for the prediction of a planet between Mars and Jupiter came largely as a result of The **Bode-Titius** progression. This numerological sequence, discovered by Titius in 1766 and popularized by Bode in 1772, predicts the distances of the planets from the Sun in terms of the Earth's distance (the astronomical unit).* The progression is constructed by writing the sequence 0, 3, 6, 12, 24..., adding 4 to each number and dividing the result by 10. Figure 4.2 shows the results of first 10 members of the Bode-Titius progression.

Bode-Titius Progression	Planet	a (au)
(0+4) ÷ 10 = 0.4	Mercury	0.39
(3+4) ÷ 10 = 0.7	Venus	0.72
(6+4) ÷ 10 = 1.0	Earth	1.00
(12+4) ÷ 10 = 1.6	Mars	1.52
(24+4) ÷ 10 = 2.8	Asteroid Belt	2.77
(48+4) ÷ 10 = 5.2	Jupiter	5.20
(96+4) ÷ 10 = 10.0	Saturn	9.54
(192+4) ÷ 10 = 19.6	Uranus	19.18
(384+4) ÷ 10 = 38.8	Neptune	30.07
(768+4) ÷ 10 = 77.2	Pluto	39.44

Figure 4.2: The Bode-Titius Progression

Notice that the progression quite accurately predicts the distances of the planets from the Sun out to Uranus, and also, that there is a prediction of a planet at a distance of 2.8 au.

In the later part of the 18th century, astronomers started a search for the "missing" planet. Therefore, when Piazzi discovered Ceres, it was expected as was the fact that it had a mean distance from the Sun of 2.8 au.

*No one knows why this progression works!

The Largest Minor Planets; The minor planets are numbered in order of their discovery and this number is followed by a name chosen by the discoverer. Since there is a general tendency to discover the larger minor planets first, they tend to have the lowest numbers. The four largest minor planets are **1 Ceres** (diameter 630 miles), **4 Vesta** (diameter 340 miles), **2 Pallas** (diameter 330 miles) and **3 Juno** (diameter 150 miles).

Altogether, there are probably some 100,000 bodies larger than one mile in diameter but only about 4,000 have been mapped and catalogued.

Minor Planets Outside the Asteroid Belt; While most of the minor planets maintain orbits which keep them between Mars and Jupiter, some have orbits which take them nearer to the Sun than Mars. The most conspicuous of these are those which form the group of **Apollo asteroids**; minor planets with Earth crossing orbits. Because of potential collisions with Earth, these bodies have aroused much interest among astronomers (Figure 4.3).

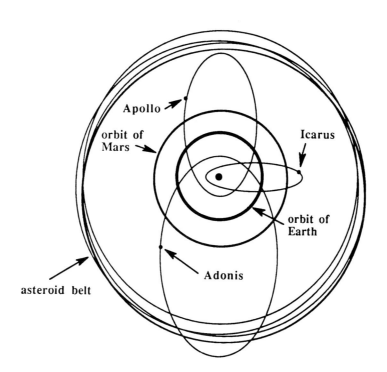

Figure 4.3: Some "Apollo" Asteroids

To date, dozens of Apollo asteroids have been catalogued. Most are hundreds of yards to several miles across. From time to time, these "boulders" collide with Earth causing great impact craters. It is hypothesized that a collision between the Earth and an Apollo asteroid, 65 million years ago, caused the extinction of the dinosaurs and about 75 percent of all other plant and animal species then existing on Earth.

It is believed that, as Apollo asteroids are swept up by Earth, more "normal" asteroids are injected into Apollo type orbits by interactions with Jupiter. In this way, the Apollo asteroids always exist.

Several dozen asteroids have been discovered at gravitationally "stable" points in the orbit of Jupiter. These stable **Lagrangian points** lie both 60° ahead of, and 60° behind Jupiter and occur as a result of the interplay of the Sun's gravity and Jupiter's gravity. Once a small body enters the "Lagrangian" zone, it is difficult for it to escape. These asteroids are called **Trojan asteroids** (Figure 4.4).

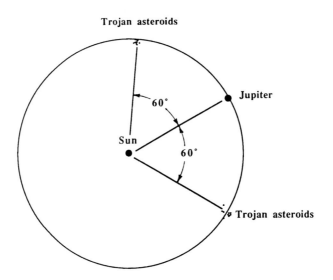

Figure 4.4: The Trojan Asteroids

The lone asteroid beyond Saturn is **2060 Chiron**. Its diameter is about 120 miles. In 1987, it suddenly brightened and developed a coma (outer atmosphere) indicating that it contained some proportion of icy material. Could Chiron be a link between asteroids and comets?

The Origin of Asteroids; When the solar system was in the process of formation, the gas and dust formed into "lumps" several miles to several hundred miles across. These lumps are called **planetesimals** and it is their accretion (joining) which created the planets. It is believed that the planetesimals in the region between Mars and Jupiter were accelerated by Jupiter's gravity and hence, when colliding, were moving too fast to "stick". Instead, they broke into fragments; the asteroids we see today.

Another effect that Jupiter has on the asteroids are the **Kirkwood gaps**; gaps in the distribution of minor planets similar to the gaps in the "ringlets" of Saturn. It is believed that the Kirkwood gaps are due to perturbations (gravitational influences) by Jupiter on those minor planets which are at particular distances from Jupiter's orbit.

The Comets; Comets are perhaps the most mysterious and beautiful members of our solar system. From time to time they are discovered silently gliding towards the Sun from some unknown point of origin beyond the orbit of Pluto. As they approach the Sun, they develop a long, graceful, vaporous tail which stretches for millions of miles and can be seen night after night, slowly changing its position against the background of more distant stars.

In ancient times, the beauty of comets was overshadowed by the belief that comets were the portents of evil; the deaths of kings, the fall of empires, the coming of famine. Halley's comet (period of 76 years) was associated with the fall of Jerusalem in 70 AD, the defeat of Attila in 451 AD and the Norman conquest of 1066 AD.

The demystification of comets came about partly as a result of the work of **Edmund Halley** in 1704. At this time, the appearance of comets was totally unpredictable; a quality which led to their affiliation with the supernatural. Halley, using mathematics developed by his close friend Isaac Newton, calculated that the comets seen in 1456, 1531, 1607 and 1682 were actually the same comet and should again be seen in 1758. When the comet reappeared as predicted, it was designated as **Halley's comet** and has remained the most famous of all of the solar system's icy celestial wanderers.

The Orbits of Comets; Most comets have orbits which are highly elliptical. They are inclined to the plane of the solar system at random angles and they orbit the Sun in all directions. Hence, they must originate from a spherical distribution. Some comets have orbits which are parabolic or hyperbolic and are thus non-periodic in that their velocities will cause them to leave the solar system forever. It is believed that interactions with the giant planets might have accelerated some comets into these "open ended" orbits. In other cases, the orbits of some comets are smaller and more circular. Interactions with planets probably decelerated these comets into "short period" orbits (Figure 4.5).

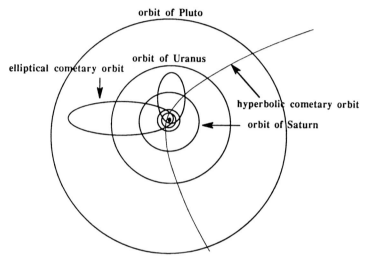

Figure 4.5: The Orbits of Comets

Most comets orbit the Sun with periods greater than 1,000 years; times so long that reappearances are not recognized as such. About 100 comets, however, have periods of less than 200 years and are called **short period comets**. Examples of short period comets are **Halley's comet** (p = 76 years) **comet Pons-Brooks** (p = 71 years) and **comet Encke** (p = 3.3 years).

The speeds of the comets in their orbits are highly variable. From Kepler's 2nd law of planetary motion (Chapter 2), we know that the farther an orbiting body is from the Sun, the more slowly it moves. We, therefore, see that comets move slowly when they are far from the Sun (near their *aphelion* point) and rapidly when they are close to the Sun (near their *perihelion* point). A comet with a highly elliptical orbit will, therefore, spend most of its time far from the Sun and only a small amount of time near the Sun where we can observe it.

The Structure of Comets; Comets appear to be divided into three parts; the **nucleus**, **coma** and **tail**. The nucleus and coma together comprise the **head**.

The nucleus of a comet may be most aptly described as a large "dirty iceberg" or "dirty snowball" moving through space. Diameters of these nuclei range from a fraction of a mile across to 10 or 20 miles in diameter. They appear to be composed chiefly of frozen water (H_2O), frozen carbon dioxide (CO_2), frozen methane (CH_4) and frozen ammonia (NH_3). Mixed in with these ices are small dust grains which are composed largely of carbon, hydrogen, oxygen and nitrogen. It is believed that some of the dust grains frozen into comet nuclei are remnants of the original material from which the solar system was formed.

During the 1986 perihelion passage of Halley's comet, the European Space Agency's comet probe (named **Giotto** after the 14th century artist who placed Halley's comet in a nativity scene) passed within 350 miles of the comet's nucleus. Photos revealed that the nucleus was an oblong body about 9 miles long by about 5 miles wide (about the size of Manhattan Island). Heat from the Sun was causing "eruptions" on the surface of the nucleus. The vapors exploding into space were mostly water vapor, with dust particles mixed in (Figure 4.6).

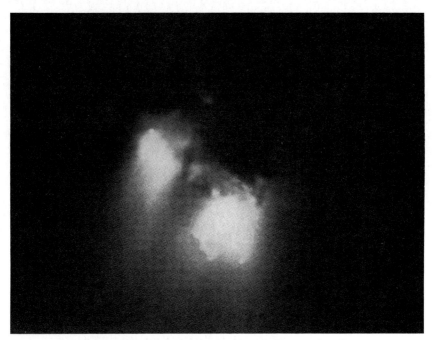

Figure 4.6: The Nucleus of Halley's Comet

When a comet is far from the Sun (as is usually the case), the nucleus is in a frozen state. As it approaches the Sun, however, some of the frozen gases in the nucleus begin to sublime (go directly from the solid state to a gaseous state). These gas molecules carry with them some of the small dust particles which were embedded in the nucleus. This gaseous, dusty outpouring forms a tenuous cloud surrounding the nucleus; the coma (Figure 4.7).

The comae of typical comets range in diameters up to 50,000 miles or more. Spectroscopic analysis of the radiation emitted by vapors in some cometary comae indicates the existence of atoms of sodium, iron, silicon, magnesium, carbon and other materials presumably ejected from the nucleus. Recently, more complex organic molecules (CH_3CN, HCO, H_2CO) have been discovered. This is particularly exciting as these materials are considered the **building blocks of life**. It is believed by some that organic material, raining down on the primeval Earth from comet infall, may have led to the origin of life on our planet.

As the comet moves still closer to the Sun (at about the distance of Mars or Jupiter), it comes into the sphere of influence of two important "solar" forces which cause the *tails* to form. The first of these forces is related to the **solar wind**, a stream of charged particles (mostly protons and electrons) ejected from the Sun. The solar wind carries with it the magnetic field of the Sun. When the solar wind comes into contact with the ionized (charged) gas atoms in the coma of the comet, these gas atoms are stripped from the coma, in a radial direction away from the Sun (Figure 4.8).

The **gas** or **ion tail** is thus formed from the interaction of the solar wind and the gas atoms of the coma. The gas tail always points away from the Sun no matter what direction the comet is moving.

Figure 4.7: The Coma of Halley's Comet; May 1910

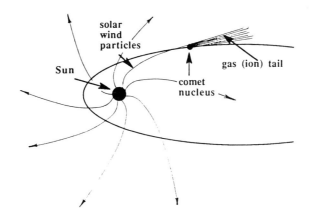

Figure 4.8: Formation of Gas Tail

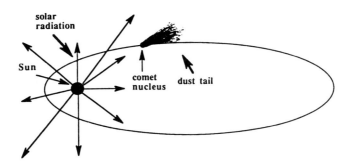

Figure 4.9: Formation of Dust Tail

A second force effecting the comet is the pressure exerted by the electromagnetic radiation (light) emitted by the Sun. This **radiation pressure** is extremely effective in pushing the small particulate matter (dust) out of the coma. The **dust tail**, then, forms in response to the pressure from the Sun's outpouring radiation (Figure 4.9).

The two tails just described both point away from the Sun but not exactly in the same direction. The gas tail is weakly curved because the magnetic field of the Sun, along which the gas atoms move, is itself curved into a spiral form by the rotation of the Sun. The dust tail is more strongly curved because, as the dust grains are driven farther and farther from the coma, they find themselves farther and farther from the Sun. This causes them to move more slowly in their orbits (in accordance with Kepler's 2nd law). They, therefore, lag farther and farther behind the coma and cause a curving of the dust tail. Comet tails are frequently as long as 100 million miles but their density is exceedingly low. The two tails of comet West are clearly seen in Figure 4.10.

Figure 4.10: Comet West; 1976

The light by which we see the two types of comet tails originates in two different manners. The gas (ion) tail shines partly by scattering blue light (as does our atmosphere) and partly by fluorescence (just as a neon gas tube gives off light when the neon is excited by an electric current). The dust tail shines by reflecting (scattering) longer wavelength visible light. The spectrum of the dust tail, rather than being representative of the dust grains, is simply a reflected solar spectrum. The evolution of comet tails is shown in Figure 4.11.

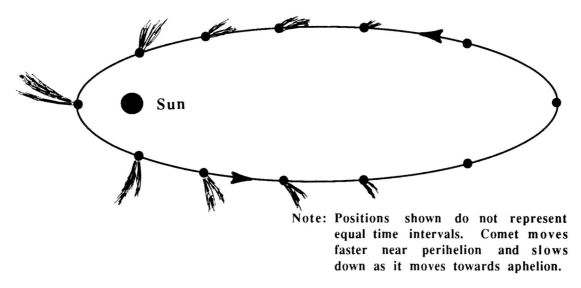

Note: Positions shown do not represent equal time intervals. Comet moves faster near perihelion and slows down as it moves towards aphelion.

Figure 4.11: The Evolution of Comet Tails

The Origin of Comets; Each time a comet passes close to the Sun, a coma and tail forms. This material is lost to space as the comet moves about the Sun. It is estimated that a comet cannot pass by the Sun more than 1,000 times before it is "used up" (some "sun-grazing" comets have even shorter lifetimes). Since comets have limited lifetimes, there must be a source of these bodies somewhere in space.

The key clues to the origin of comets comes from their spherical distribution about the plane of the solar system and the highly elliptical nature of their orbits. In 1950, the Dutch astronomer Jan Oort studied the orbits of hundreds of comets and found that they spend most of their time at a distance of 50,000 to 100,000 au from the Sun. He hypothesized a "cloud" of comet nuclei, **The Oort cloud**, centered on the Sun and containing, perhaps, millions of comets. At this distance, the cometary nuclei are so far from the Sun that they cannot be seen by earthbound observers. Periodically, however, these icy nuclei are disturbed by the passage of a nearby star or by interactions with one another. At these times, some of their orbits are changed so that they fall in towards the inner solar system and the Sun. They become visible as the coma and the tails form. Sometimes, however, interactions within the Oort Cloud will eject the comet nucleus into interstellar space and the comet will be lost to the solar system forever (Figure 4.12).

The Naming of Comets; The honor of naming a newly discovered comet (about 10- 20 "new" ones are found each year) goes to the discoverer of the comet, generally astronomers who accidentally photograph them while studying other celestial objects. Frequently, comets are discovered by amateurs who sweep the skies each clear night with wide angle telescopes or binoculars.

In addition to the comet's "proper" name (e.g. Comet West, Comet Bennett, Comet Kohoutek), it is given a temporary designation based on its year of discovery and order of discovery within that year. Therefore, the comets discovered in 1992 are 1992a, 1992b, 1992c...

etc. The permanent astronomical designation of the comet consists of the year in which it passed perihelion (closest approach to the Sun) followed by a Roman numeral indicating its order of perihelion passage in that year. For example, the first comet to reach perihelion in 1992 will be permanently designated as comet 1992 I.

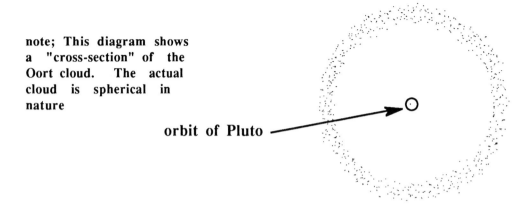

note; This diagram shows a "cross-section" of the Oort cloud. The actual cloud is spherical in nature

orbit of Pluto

Figure 4.12: The Oort Cloud

Meteors and Meteoroids; Unlike comets, meteors are fleeting phenomena. A comet, orbiting the Sun millions of miles from the Earth, may be seen for weeks on end, only slowly changing its position in the sky. Meteors, on the other hand, are formed when small grains or particles enter the Earth's atmosphere at high speeds (up to 45 miles/second) and burn up due to friction with the atmosphere. The flash of light due to such an event is called a **meteor**. The particle which causes the meteor phenomenon, when floating in space, is called a **meteoroid**. On the other hand, if the particle survives its trip through the atmosphere and reaches the Earth, it is called a **meteorite**.

On any clear, dark night an observer can see from 5 to 10 **sporadic meteors** per hour. These meteors occur at an average height of about 50 miles above the surface of the Earth and may be seen coming from any direction (they show no preference for the ecliptic plane). On any night, more meteors are seen during the early morning hours (2:00 am to 5:00 am) than before. This is due to the fact that, during these hours, the observer is on the leading edge of the Earth; the side which is rushing forward through space. It is this side of the Earth which "sweeps up" many small particles. The trailing edge of the Earth tends to run away from particles in space (Figure 4.13).

Exceptionally bright meteors are called **fireballs** or **bolides**. They may be bright because the meteoroids which cause them are large or because they come into the Earth's atmosphere relatively slowly and, therefore, survive to rather low altitudes before they vaporize.

On certain nights each year, the hourly meteor rate increases dramatically. These **meteor showers** occur when the Earth moves into a region of space where the particle density is higher than average. Such high density regions are found along the orbits of living or defunct comets (defunct comets are those which have been shredded by the Sun or have ejected most or all of their icy material). If the Earth passes through the orbit of such a comet, the particles in the orbit cause an enhancement of the rate of meteor occurrence (Figure 4.14).

Associated with each meteor shower is a so-called **radiant point**; a particular point in the sky from which the meteors of a particular shower appear to emanate. The radiant point of the meteor shower which peaks on October 20th each year is shown in Figure 4.15. The reason for the radiant point is shown in Figure 4.16. As one looks down a long straight road it is apparent that the sides of the road, the telephone poles and the trees along side the road all appear to converge to a point in the distance. This is what is called perspective. Similarly, if two motorcycles were riding towards you along opposite roadsides, they would appear to diverge from the same point.

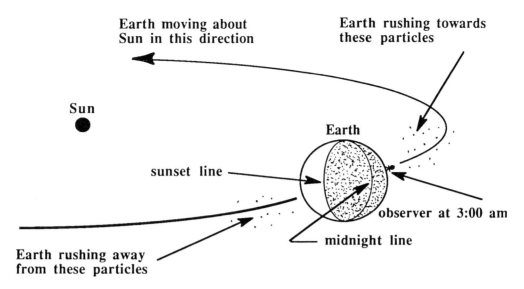

Figure 4.13: Formation of Meteors

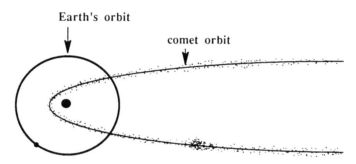

Figure 4.14: Intersection of Orbits of Earth and Comet

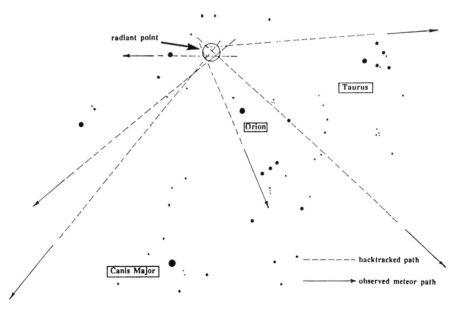

Figure 4.15: The Orionid Meteor Shower

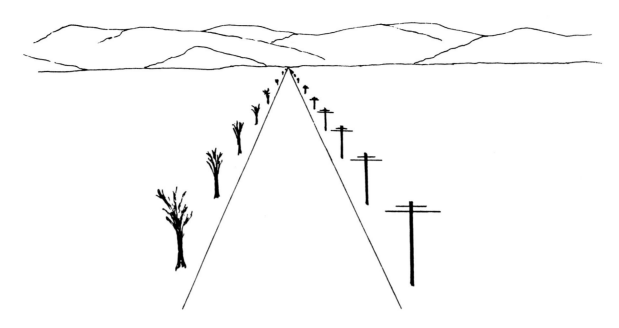

Figure 4.16: Perspective

The particles which cause the shower meteors are moving along parallel paths about the Sun (Figure 4.17). Therefore, when we observe their paths (meteor flash), they appear to diverge from a point in the sky (radiant point). The point actually represents the direction from which we see the particles approaching (Figure 4.15).

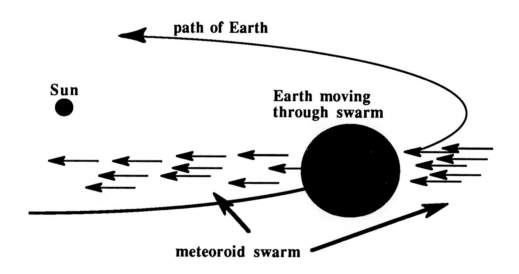

Figure 4.17: Orbit of Earth and Particle Swarm

Each meteor shower is named for the constellation in which the radiant point falls (e.g. Orionid shower) or for a bright star near the radiant point (e.g. Eta Aquarid shower). The best meteor showers are given in Figure 4.18 (Learning Activity 11, Appendix IV).

Shower	Date of Maximum	Associated Comet	Approximate Hourly Rate
Quadrantids	January 3	?	40
Lyrids	April 21	1861 I	12
Eta Aquarids	May 4	Halley	20
Delta Aquarids	July 30	?	35
Perseids	August 12	1862 III	68
Draconids	October 9	Giacobini-Zinner	15
Orionids	October 20	Halley b	30
Taurids	October 31	Encke	12
Leonids	November 16	1866 I	10
Geminids	December 13	?	58

Figure 4.18: Meteor Showers

Meteorites; When meteoroids pass through the Earth's atmosphere at high speed, they generally burn up, resulting in the phenomenon known as a **meteor**. From time to time, however, these rocks from space survive to the surface of the Earth where they can be found and studied. On Earth, these meteoroids are called **meteorites**. Prior to the 19th century, reports that rocks could fall from the sky were treated with skepticism and derision. It was not until a wide area of France was struck by fragments of an exploded meteoroid in 1803, that eyewitness accounts rendered the phenomenon irrefutable.

It is now realized that the Earth is constantly being bombarded by "rocks" from space. These rocks range in size from microscopic ("micrometeorites") to miles across. Obviously, the small impacts occur far more frequently that the large impacts. While many thousands of tons of micro meteorite material fall to Earth each day, large boulders, hundreds of feet in diameter, strike the Earth once or twice in thousands of years.

The most famous meteoroid impact site in the United States is **Meteor Crater** near Flagstaff, Arizona. The crater was created when a 100 foot diameter meteoroid struck the Earth (Figure 4.19). The event occurred sometime between 20,000 and 50,000 years ago. The crater which resulted is almost a mile in diameter and 600 feet deep!

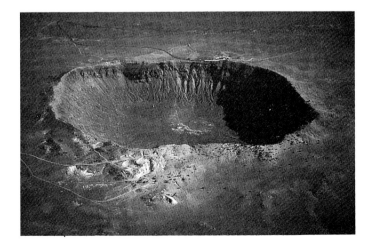

Figure 4.19: "Meteor Crater"

One of the strangest events associated with the infall of meteoritic material occurred on November 30, 1954 in Sylacauga, Alabama, when a nine pound iron meteorite crashed through the roof of a house, bounced off a bedstand and hit a sleeping woman in the hip causing a severe bruise. Ermeline Hewlett Hodges is the only person known to have been struck by a meteorite from space (Figure 4.20).

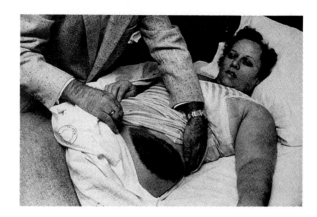

Figure 4.20: Ermeline Hewlett Hodges

Meteorite Types; The meteorites which are found on Earth appear to originate from material in the asteroid belt. The meteorites appear to be divided into three major types:
- **Stony meteorites**: These objects appear to be samples of the crustal material of the parent bodies (asteroids) from which these meteorites originated. They come in two varieties; **chondrites** and **achondrites**. **Chondrites** (so-called because they contain spherical glassy BB's called **chondrules**) appear to be samples of material created at the time of the formation of the solar system and not significantly altered since then. The most interesting type of chondrite, the **carbonaceous chondrites**, are stone meteorites which contain no chondrules and are similar to terrestrial basalts (solidified lava). Altogether, the stony meteorites comprise about 93 percent of meteorites observed to fall. This indicates that over 90 percent of the meteoroid material in space is of the stony variety. The largest stone meteorite observed to fall on Earth was a two ton mass which fell on Kirin Provence in China on March 8, 1976.
- **Iron meteorites**: These objects are composed of nickel-iron and are thought to be samples of the cores of the parent asteroidal bodies. When large asteroids formed coalescing planetesimals (Chapter 7), they became molten from the heat liberated by radioactive material. Heavier material (iron, nickel, etc) settled to the centers of these asteroids and formed metallic cores. When the asteroids were shattered by collisions, pieces of the solidified cores became iron meteoroids. Iron meteorites account for six percent of observed meteorite falls and it was an iron meteorite which created **Meteor Crater** in Arizona. The largest iron meteorite discovered is a 50 ton chunk at Hoba West in South Africa. The largest in a museum is a 35 ton object, found by Peary in Greenland and located at the Hayden Planetarium of the American Museum of Natural History in New York City.
- **Stony-iron meteorites**: These meteorites are mixtures of stone and iron and are probably samples of the interface between the cores and crusts of the parent bodies. Only one to two percent of the observed falls are stony-iron meteorites.

Radiometric dating techniques indicate that meteorites formed about 4.6 billion years ago and are, therefore, samples of the most primitive solar system material. They are, therefore, useful in our studies relating to the origin of the Solar System.

CHAPTER 5
THE MOON

The Moon is the Earth's nearest neighbor in space and the celestial body about which most is known. It is the only body in the solar system (aside from Earth) on whose surface humans have walked and the only body (aside from, perhaps, Mars) from which we have soil samples. Because the Moon is so close to Earth, it has a strong gravitational effect on our planet and its tidal action may have played an important part in the evolution of life on Earth.

The Moon is also the most interesting object for naked eye observation. It's motion against the stellar background is the most rapid of any celestial body and it is the only object which shows visible surface features and periodic variations in illumination (its phases).

The Orbit of The Moon; The Moon orbits the Earth in an elliptical path (as predicted by Kepler's First Law) at an average distance of 239,000 miles. Because of the eccentricity of the orbit, however, the distance varies from 226,000 miles (perigee) to 252,000 miles (apogee).* When the Moon is at perigee it appears slightly larger in the sky than when it is at apogee. This will become important when we discuss eclipses. The orbit of the Moon is shown in Figure 5.1.

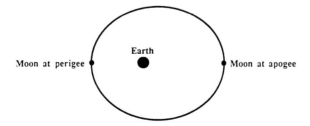

Figure 5.1: The Orbit of the Moon

*Perigee and apogee refer to an Earth satellite's nearest and farthest distances from Earth.

The Month; The period of revolution of the Moon about the Earth is called the **month**. The exact duration of the month, however, depends on what one chooses as a point of reference. The two most commonly used "months" are the **synodic month**, which uses the Sun as a point of reference and the **sidereal month**, which uses the stars as a point of reference. The difference between these two months is shown in Figure 5.2.

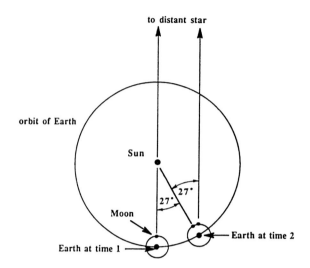

Figure 5.2: The Sidereal and Synodic Month

From the diagram, you can see that at time 1 (t_1) the Moon is between the Earth and the Sun. Also, at t_1, the Moon is between the Earth and a distant star. At t_1, we will let both the synodic and sidereal month begin. As time goes by, the Moon revolves about the Earth while the Earth revolves about the Sun. At t_2, the Moon has moved back to a position between the Earth and <u>the same star</u> (since the stars are very far away, observers relatively near one another will look along parallel lines to see a given star). The time interval from t_1 to t_2 is 27d 7h and is defined as the sidereal month (the period of revolution of the Moon with respect to the stars). But the synodic month has not yet been completed because the Moon has not yet moved back to a position between the Earth and the Sun. For this to occur it takes an additional 2d 5h (it takes this long for the Moon to move 27° around the Earth. The synodic month, therefore, is 29d 12h in duration. The phases of the Moon repeat every synodic month.

The Orbit Plane of the Moon; The plane of the Earth's orbit about the Sun is called **the ecliptic plane** and the path of the Earth about the Sun is called **the ecliptic**. This is shown in Figure 5.3.

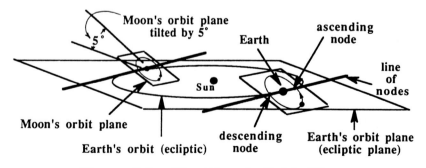

Figure 5.3: The Moon's Orbit Plane

The orbit plane of the Moon is inclined by about 5° to the plane of the ecliptic. The points where the Moon's orbit crosses the ecliptic are called **nodal points**. The point where the Moon crosses from south of the ecliptic to north of the ecliptic is called the **ascending node** and the point where the Moon crosses from north of the ecliptic to south of the ecliptic is called the **descending node**. The line on the ecliptic plane, connecting the two nodes is called the **line of nodes**.

Because of gravitational forces from the Sun and planets acting on the Moon, the orbit plane of the Moon is twisting in space with a period of 18.6 years. This twisting is causing the nodes to swing westward (clockwise when viewed from the north) and is called the **regression of the line of nodes**. The regression of the line of nodes will become important in our discussion of eclipses (Figure 5.4).

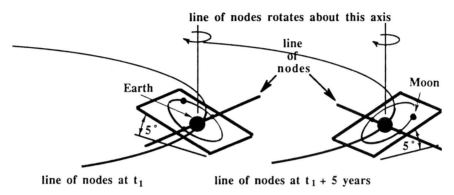

Figure 5.4: The Regression of the Line of Nodes

The Phases of the Moon; The progression of lunar phases is the most dynamic phenomenon of the night sky and is familiar to all. Despite this, however, few people understand its cause. Simply stated, the phases of the Moon progress as one is able to see first more and more and then less and less of the illuminated half of the Moon (Figure 5.5).

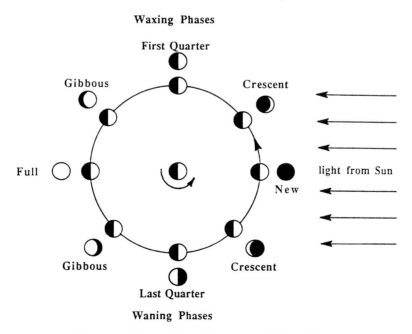

Figure 5.5: The Phases of the Moon

It must first be realized that the Moon emits no light of its own but rather reflects sunlight (the Moon reflects about 7% of the incident sunlight; its **albedo** is 7%). When the Moon, in its revolution about the Earth, takes a position between the Earth and the Sun*, the lit (illuminated) side of the Moon is turned away from the Earth and we see nothing when we look towards the Moon (note that we are also looking into a daytime sky). This lunar phase is called **new moon** and it occurs every synodic month (29d 12h).

For the next 15 days, the Moon progressively moves around the Earth (at a rate of about 12°/day) and as it does so, an earthbound observer will see more and more of the lit side. Starting from day 0 (new moon), the first lunar phase is the **crescent moon** (day 1 to day 6). Figure 5.6 shows a 3 day and 5 day old crescent moon. Notice that a crescent moon is less than half illuminated because the lit side of the Moon is still turned mostly away from the Earth.

Figure 5.6: A 5-Day Old and 3-Day Old Crescent Moon

On day 7, the Moon reaches the 1/4 mark in its trip about the Earth and an earthbound observer sees half of the illuminated half (Figure 5.7). This is the **first quarter moon**.

For the next 7 days, more than half of the illuminated half can be seen (Figure 5.8). This is called the **gibbous moon**.

On day 15, the Moon finds itself halfway around in its orbit, on the other side of the Earth from the Sun. In this position, the full illuminated lunar hemisphere is facing the Earth and observers see a **full moon** (Figure 5.9).

The phases from new to full moon (during the first 15 days of the lunar cycle) are collectively known as **waxing phases**. Note that during the waxing phases, an observer will see the right side of the Moon illuminated. The reverse is true during the next 15 days.

From day 16 to day 30, as the Moon moves back towards the Sun, the phases repeat in reverse order going from **full** to **gibbous** to **last quarter** to **crescent** and finally to **new**. The phases from full to new are called **waning phases**. Notice that during the waning phases, an observer will see the left side of the Moon illuminated.

*Actually, the Moon is <u>directly</u> between the Earth and the Sun only 2 to 5 times a year. At other times, its orbit passes slightly lower or slightly higher than the Earth-Sun line.

Figure 5.7: The First Quarter Moon

Figure 5.8: The Gibbous Moon

Figure 5.9: The Full Moon

An interesting aspect of the lunar phases is that they are always in the same part of the sky at the same time of day. Notice from Figure 5.5 that the new moon is in the same direction as the Sun and therefore must rise at sunrise and set at sunset. The full moon, since it is opposite the Sun, must rise at sunset and set at sunrise. The first quarter moon, since it is halfway between new and full moon, must rise at noon and set at midnight. Can you determine, from Figure 5.5, the rising and setting time of the last quarter moon (Learning Activity 12, Appendix IV)?

Another important aspect of the lunar phases relates to the daily delay in moonrise (Figure 5.10).

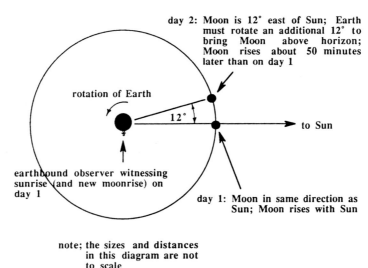

Figure 5.10: The Delay in Moonrise

Assume that on "day 1" the Moon is new and rises at sunrise. One day later (day 2), the Moon has moved about 1/30 of the way (12°) around the Earth (relative to the Sun). Since the revolution of the Moon and the rotation of the Earth are in the same direction, the Earth must turn an additional 12° on "day 2" to bring the Moon above the horizon. The Earth takes 50 minutes to rotate 12° and, therefore, the Moon will rise 50 minutes later on "day 2" than on "day 1". Each successive day, the Moon rises about 50 minutes later than the day before. The number 50 minutes is only an average. The actual amount depends on the time of year and your location on Earth.

The Distance to the Moon; As already stated, the distance to the Moon averages 239,000 miles. Up until 40 years ago, the only way of measuring this distance was by triangulation. This method, which is over 2000 years old, involves observing the Moon from two widely separated locations (Figure 5.11) and noting the relative difference in the lunar direction (angle "P" in Figure 5.11). From a knowledge of this angle and the distance between the 2 observing stations (distance "B" in Figure 5.11), the distance to the Moon can be calculated. The equation used for this calculation is **Distance = B/P** where the units of distance and the units of B are the same and angle P is in radians*.

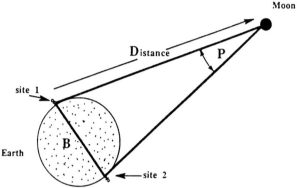

Figure 5.11: Triangulation

In the 1950's a new method was developed to measure the lunar distance. The method involves bouncing radio waves off the Moon and timing the interval from their transmission to their reception back on Earth. Since we know the speed of radio waves (186,000 miles/second), the distance to the Moon can be calculated. This method, known as **radar**, yields an accuracy of about 1 mile.

A more accurate variation of the radar method was developed in 1969 when the Apollo astronauts placed specially designed **laser reflectors** on the lunar surface. Laser beams, transmitted from observatories in Hawaii and Texas, reflect off the reflectors and return to Earth, where they are received by telescopes. Measuring round trip time and knowing the speed of light, the distance to the Moon can be calculated. Although this method uses the same principle as the radar method, it yields more accurate results because it is possible to time the round trip of a beam of laser light more accurately than the round trip time of a beam of radio waves. This method yields an accuracy of less than 6 inches!

The Physical Properties of the Moon; Many of the physical properties of the Moon are easily observable from the Earth and have been known for decades or even centuries. Other characteristics, however, require samples of lunar surface material or the placement of instruments on the lunar surface. In this section we will examine the lunar properties which can be determined by earthbound observation.

*A "radian" is a unit of angle which equals 57.3 degrees.

The Rotation of the Moon; It is common knowledge that the Moon keeps one face turned perpetually towards the Earth and, because of this, it is assumed that the Moon does not rotate. Actually, the reverse is true; if the Moon keeps the same face turned perpetually towards the Earth, it must rotate! Figure 5.12 shows the hypothetical situation in which the Moon, as it orbits the Earth, does not rotate.

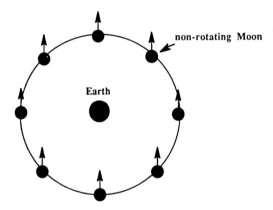
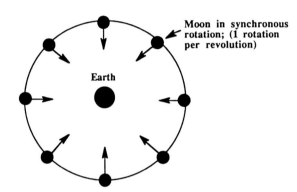

Figure 5.12: Non-rotating Moon   Figure 5.13: Synchronous Rotation

If, at point 1, we bury an arrow in the soil of the Moon pointing upwards towards the Earth, the arrow will continue to point upwards as the Moon revolves. An observer on the Earth will therefore see all sides of the Moon during a period of one month. On the other hand, if the Moon rotates once as it revolves about the Earth, the arrow will continue to point at the Earth (Figure 5.13) and an earthbound observer will only see one side of the Moon.

This situation is called **synchronous rotation**. The sidereal period of rotation of the Moon is, therefore, 27d 7h; the same as the sidereal period of revolution.

The synchronous rotation of the Moon is no accident. It results from the tidal interaction of the Earth and the Moon. The Moon is not a sphere but, rather, has a slight bulge on one side. Over billions of years, the gravity of the Earth has acted on the bulge to force it to point towards the Earth. The Earth has "captured" one side of the Moon. This effect is common in the solar system; most of the major satellites of the planets are in synchronous rotation. More will be said about synchronous rotation in the section on tides.

The Diameter of the Moon; The diameter of the Moon is approximately 2000 miles. This diameter may be readily calculated once the distance and angular size of the Moon are known. The geometry of this calculation is shown in Figure 5.14.

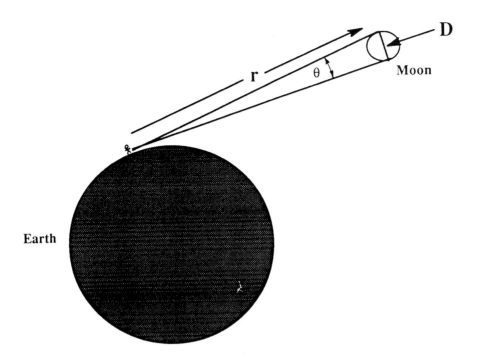

Figure 5.14: Determining the Diameter of the Moon

From the triangle in the above diagram, one can write the expression:

D = r x θ   where **D** is the diameter of the Moon
r is the distance to the Moon
and   θ is the angular size of the Moon in radians
(1 radian = 57.3 degrees)

Since we know that r = 239,000 miles and we can measure θ to be about 1/2 degree, the diameter of the Moon can be readily calculated. This diameter is about 2000 miles or 1/4 Earth's diameter.

The Mass of the Moon; Whenever two bodies are in mutual revolution (e.g., the Earth and the Moon), *both* bodies actually revolve about the center of mass (average mass point) of the system. The center of mass is the balance point of the system. For example, if the Earth and the Moon had the same mass, the center of mass would be halfway between them and it would be about this point that each object revolved. If the Earth were twice as massive as the Moon, each body would revolve about a point which was twice as close to the center of the Earth as to the center of the Moon. This means that we can determine how much more massive the Earth is than the Moon by determining where the center of mass (called the **barycenter**) of the system lies. How can we do this? We can do this if we remember that not only does the Moon orbit the barycenter, but the center of the Earth does also. The center of the Earth moves about this point each month. This motion of the Earth shows up as a monthly displacement of the nearby planets (e.g., Mars). The magnitude of this displacement indicates how far the center of the Earth is from the barycenter. Experiment shows the barycenter to be about 3000 miles from the center of the Earth. The center of the Moon is about 240,000 miles from the barycenter. The barycenter is, therefore, roughly 80 times farther from the center of the Moon than from the center of the Earth. This indicates that the Moon is about 1/80 the mass of the Earth (the exact number is 1/81.3). This is illustrated in Figure 5.15.

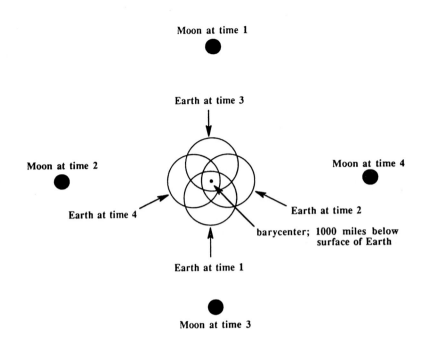

Figure 5.15: Determining the Mass of Moon

Figure 5.16: Galileo's Sketches of the Moon

    The Lunar Surface; The Moon is the only celestial body which shows surface features to the naked eye. The details of the lunar surface, however, could not become known until after the invention of the telescope in about 1608. In 1609 and 1610, Galileo Galilae made the first recorded observations of the Moon (Figure 5.16). He reported that the Moon had a rough, uneven surface with mountains, chasms and craters. He found the dark gray regions, which were visible to the naked eye, to be large, relatively flat areas which he mistook for seas (**maria** in Latin). It was later realized that these "seas" were actually large flat planes, but the name stuck. There are 14 seas on the near side of the Moon. The largest being **Mare** (pronounced *mah' ray*) **Imbrium**, the Sea of Showers, which is about 700 miles across (Figure 5.17). The major lunar seas are shown in Figure 5.18. Rocks returned from Apollo Moon missions show the maria material to be similar to basaltic lava flows. These dark rocks give the seas their characteristic dark color. Radiometric dating places the lunar seas between 3.1 and 3.9 billion years old.

Figure 5.17: Mare Imbrium Region of the Moon

Apart from the lunar seas, the most obvious lunar features are the **craters**. Even with primitive telescopes, it was possible to see thousands of impact craters; circular depressions caused by meteoric bombardment early in the history of the Moon. The largest craters are about 150 miles in diameter and the smallest are microscopic. In 1650, an Italian priest, Giovanni Riccioli, initiated the modern practice of naming the lunar craters after scientists and philosophers (Figure 5.18 and 5.19).

The lunar surface is roughly divided into two types of terrain; heavily cratered, light colored **lunar highlands** and relatively smooth, dark **lunar seas** or **lowlands**. The exact physical nature of these regions could not be determined until lunar rock samples were brought back from the Moon by Apollo astronauts. Highland rocks are a type know as **breccias** (meaning broken) which consist of fragmented and cemented material, obviously the result of impacts and heating.

Around the perimeter of some lunar seas are mountain ranges which reach heights of 25,000 feet. These ranges are formed by debris presumably created during the impact event which formed the basin which became the lunar sea. These ranges are named after mountain ranges on Earth (Figure 5.20).

Figure 5.18a: The Western Hemisphere of the Moon

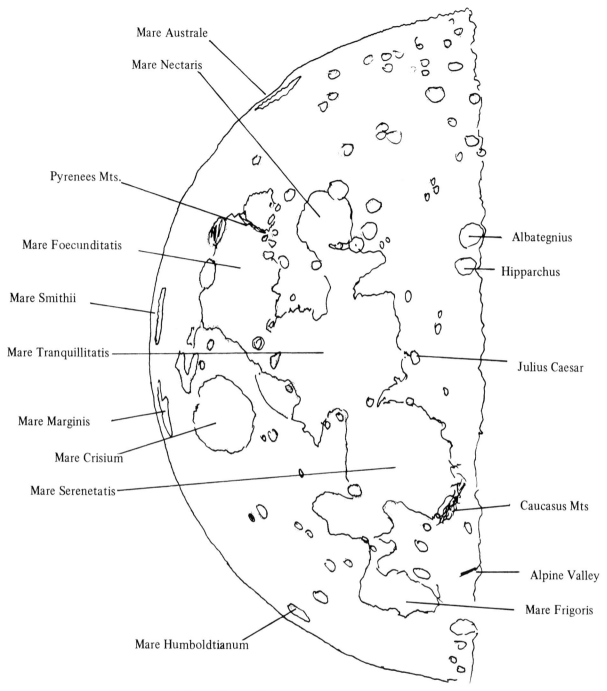

Figure 5.18b: The Western Hemisphere of the Moon

Figure 5.18c: The Eastern Hemisphere of the Moon

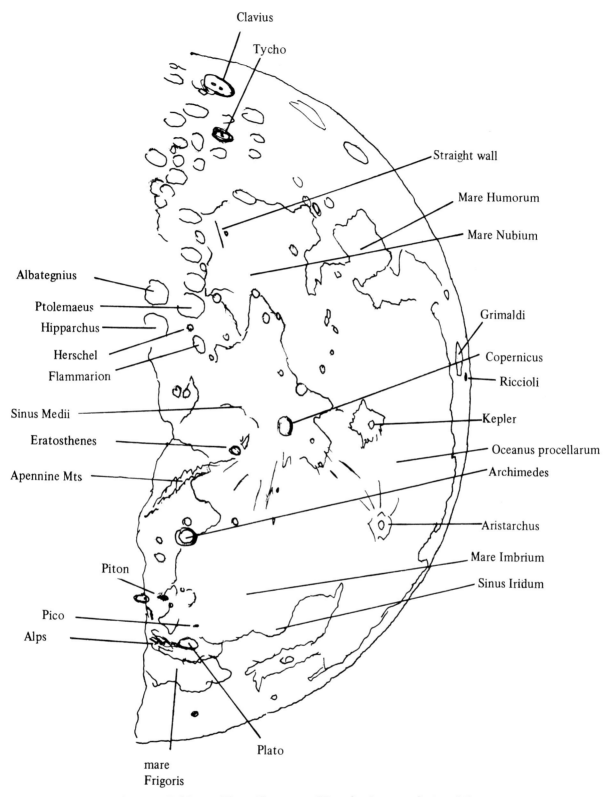

Figure 5.18d: The Eastern Hemisphere of the Moon

Figure 5.19: Lunar Craters: Region of Clavius

Figure 5.20: The Lunar Mountain Ranges Bordering Mare Serenetatis

Meandering across the lunar surface for up to hundreds of miles are sinuous depressions called **rilles** (Figure 5.21). These rilles might mark regions where the lunar surface has weakened and collapsed.

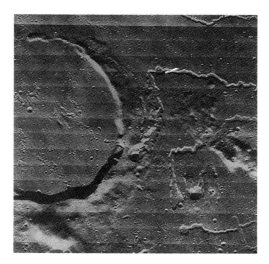

Figure 5.21: Lunar Rilles and the Crater Aristarchus

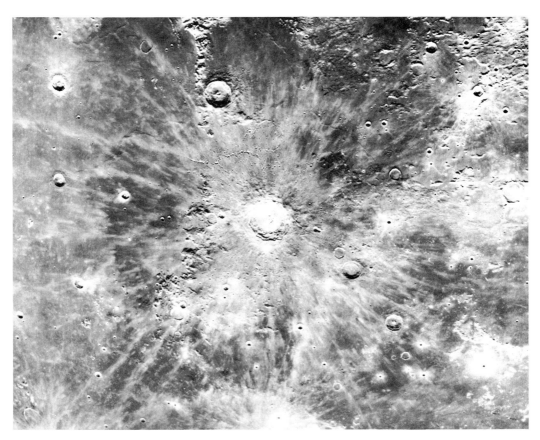

Figure 5.22: Lunar Rays Surrounding Copernicus

Radiating from many craters are bright streaks called **rays** (Figure 5.22). These rays consist of light colored lunar sub-surface material thrown outward at the time of the impact which formed the crater. As time goes by, the rays darken and disappear. It is found that rays accompany only the youngest craters.

The Lunar Surface Gravity; As seen in Chapter 2, the force of gravity between two objects is directly proportional to the product of their masses and inversely proportional to the square of their separation. Since the mass of the Moon is 1/81 the mass of the Earth, we might expect its surface gravity to be less than the Earth's in this ratio. However, since the Moon is smaller than the Earth, a person standing on the Moon is closer to the center of gravity than one standing on the Earth. The Moon's radius is 1/4 that of Earth and hence its surface gravity is;

$$\frac{1/81}{(1/4)^2} = 1/6 \text{ the Earth's gravity}$$

This lower surface gravity for the Moon leads to a lower escape velocity for this body. While the escape velocity for the Earth is about 7 miles/second, the figure for the Moon is close to 1.5 miles/second. This has important consequences when discussing the lunar atmosphere (or lack thereof).

The Lunar Atmosphere; For years, astronomers have known that the Moon has no atmosphere. This deficiency may be detected from the surface of Earth. One way is to observe the occultation (eclipsing) of stars and planets by the lunar disk. The fact that these occultations are instantaneous (rather than gradual) is indicative of an airless Moon. It is also observed that the lunar **terminator** (day-night line) is sharp and distinct. On Earth, where there is an atmosphere, the day-night line is broad and "fuzzy"; the so-called **twilight zone**. A sharp terminator (the lack of a twilight zone) is evidence for an airless world.

Once that we have established that the Moon has no atmosphere, the question which remains is why this is so. The answer follows once we understand that the presence of a planetary atmosphere hinges on two factors. One factor is the surface gravity of the body (or more precisely, its escape velocity*). The second factor is the surface temperature of the body because the speed of molecules in a planetary atmosphere is represented by the surface temperature. If the temperature of the planetary surface is such that the gas molecules can exceed the escape velocity, the molecules will escape. This is the case on the Moon and hence it has no atmosphere. It should be noted that not all molecules travel at the same velocity. More massive molecules travel more slowly than do less massive molecules (hydrogen is less massive than oxygen and moves more rapidly at any given temperature). On Earth, hydrogen and helium molecules exceed the escape velocity and hence are not found in our atmosphere. Oxygen and nitrogen do not exceed the escape velocity and therefore are present.

Water on the Lunar Surface; If the Moon has no gaseous atmosphere, it is unlikely that liquid water could exist on its surface. At the surface of the Earth, the weight of our atmosphere retards the evaporation or boiling (rapid evaporation) of water. The atmosphere "weighs down" on each square inch of surface with a force of about 15 pounds. The less weight of atmosphere there is, the more easily water will evaporate or boil. This weight is called **air pressure**.

At the Moon's surface, where there is no atmosphere, water will rapidly evaporate at <u>any</u> temperature above freezing. The water molecules would then, because of the low escape velocity, escape into space.

It is possible that water might exist in the liquid or frozen state beneath the surface of the Moon where it would not be exposed to the vacuum of space. No proof for this speculation presently exists.

*"Escape velocity" is the velocity required to escape a gravitational body.

The Lunar Surface Temperature; Because the Moon and the Earth are both about the same distance from the Sun, they both receive about the same amount of solar radiation. This might make one suspect that they should both be about the same temperature. This is, however, not the case.

The atmosphere of the Earth acts as a protective blanket; absorbing heat and, therefore, shielding us from the Sun during the day; releasing that heat and thereby warming us at night. The oceans of the Earth serve the same function. Because of these factors, the day-to-night temperature variation on the Earth's surface is seldom more than 40 to 50 F°.

Since the Moon has neither atmosphere nor oceans, it has no protection from the direct solar rays. Also, since its rotation period is a month long, the Sun is above the horizon for two weeks at a time. These two factors combine to push daytime temperatures to 265° F. During the long lunar night, the heat from the surface rapidly radiates into space and the temperature plunges to below -175° F. The day-to-night temperature swing on the Moon is therefore in the neighborhood of 450 F°.

The Lunar Space Program; On October 4, 1957, the space age began with the launch of Sputnik I. Americans were shocked to find themselves behind the Soviet Union in satellite technology and a frantic push was initiated to catch up. To galvanize America for the technological battle, President Kennedy, in 1961, committed his government to placing a man on the Moon by the end of the decade (he also thought it might be a good idea to get him back alive).

During the 1960's, Americans excitedly watched as rocket after rocket was launched into space; each rocket bringing us closer to a manned lunar landing. The American Moon program consisted of a two pronged attack; an unmanned program and a manned program. The unmanned program, directed from the Jet Propulsion Laboratory in Pasadena, California, consisted of three series of missions.

The Ranger Series; In 1964 and 1965, three **Ranger** spacecraft made successful trips to the Moon, sending back pictures until they purposely crashed into the lunar surface. The best of these pictures had a resolution of about one foot (Figure 5.23).

Figure 5.23: Ranger Photo of Lunar Surface

Although Ranger allowed us to perfect our rocket guidance systems, it did not help us to perfect soft landing techniques and did not give us much information about the general nature of the lunar surface.

The Surveyor Series; **Surveyor** was the first American lunar **soft-lander**. In 1966 and 1967, five Surveyors landed on the Moon. These craft carried T.V. cameras and soil testing experiments. The Surveyors gave us our first view from the surface of the Moon and demonstrated that the lunar surface could support a load of high mass.

Figure 5.24: (Top) Surveyor 7 Photo of Lunar Surface
(Bottom) Surveyor 3 Being Inspected by Apollo 12 Astronaut Pete Conrad

Figure 5.25: Lunar Orbiter Photo of Alpine Valley

Figure 5.26: Lunar Orbiter Photo of Earthrise

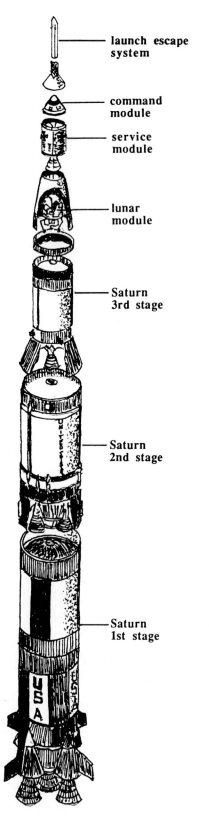

Figure 5.27: The Saturn V Launch Vehicle

Figure 5.24 (top) is a photo taken by Surveyor 7 in January, 1968. The trenches were dug by the spacecraft's surface sampler mechanism. Figure 5.24 (bottom) shows Apollo 12 astronaut Pete Conrad inspecting Surveyor 3 (note the lunar module in the background).

The Lunar Orbiter Series; In 1966 and 1967, five **Lunar Orbiter** spacecraft were sent to the Moon. As their name suggests, these craft were put into lunar orbit from where they were able to completely photograph the lunar surface. From these photos, possible Apollo landing sites were chosen (Figures 5.25, 5.26).

At the same time as the unmanned program was being carried out, a manned lunar program was being directed from the **Manned Space Flight Center** in Houston, Texas. The manned program was designed to sharpen our abilities to place humans in space, keep them alive, and return them to Earth. The manned program was divided into three parts:

The Mercury Series; Early in the manned space program, the United States had difficulty putting large payloads into orbit. The problem was building large enough rockets. Our first attempt at putting a human into orbit was in a capsule large enough for a single person; the Mercury capsule. **Alan Shepard** was the first to ride a **Mercury capsule** into space and **John Glenn** was the first to be put into orbit.

The Gemini Series; Following Mercury, the **Gemini capsules** carried 2 astronauts into Earth orbit. Once there, experiments were carried out to see how humans survived weightlessness for prolonged periods of time. It was during a Gemini mission that Charles White made the first **spacewalk**. White left the capsule and, while tethered with a cable, fell around the Earth at more than 17,000 miles per hour.

The Apollo Series; The **Apollo series** was designed to carry three astronauts from the Earth to the Moon. The Apollo capsule rode into space atop the Saturn V launch vehicle which also carried a small craft, the **lunar module**, designed to descend to the lunar surface (Figure 5.27).

On July 16, 1969, **Neil Armstrong**, **"Buzz" Aldrin** and **Michael Collins** were launched into Earth orbit by the Saturn V. The command module housing the astronauts attached itself to the lunar module which was folded inside the third stage of the launch vehicle. The engines of the service module were fired and the astronauts started their three-day journey to the Moon (Figure 5.28).

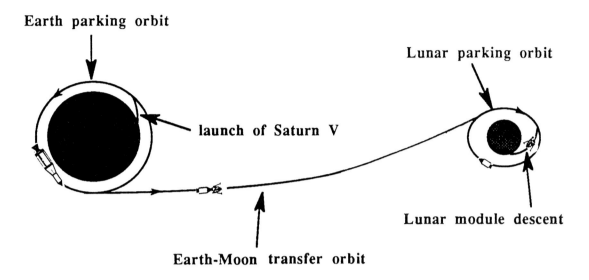

Figure 5.28: Travelling to the Moon

On July 20, 1969, Neil Armstrong and "Buzz" Aldrin transferred from the command module to the lunar module while Michael Collins remained in orbit aboard the command module **Columbia**. Armstrong and Aldrin used the engines of the lunar module **Eagle** to slow the vehicle's orbital motion and to descend to the lunar surface (Figure 5.29).

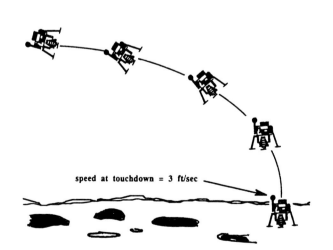

Figure 5.29: Descending to the the Lunar Surface

Figure 5.30: Blasting Off from the Lunar Surface

On board the lunar module were experiments designed to enhance our understanding of the Sun and Moon including: a foil, deployed on the lunar surface, with which to capture "solar wind" particles from the Sun; a seismometer designed to measure "moonquakes" and vibrations from meteoroid impacts and thus probe the lunar interior; a specially designed laser reflector which would allow astronomers to measure the Earth-Moon distance to an accuracy of less than six inches; and a core sampler which would allow samples of lunar subsurface material to be returned to Earth. Additionally, Apollo teams returned about 1,000 pounds of moonrocks to Earth. These rocks have allowed us to learn much about the origin and structure of the Moon. In all, six Apollo teams visited the lunar surface from 1969 to 1972. The last Apollo missions even carried electric cars (**lunar rovers**) to allow astronauts to explore at greater distances from the lunar module.

When the two-man Apollo crew was ready to leave the lunar surface and rejoin their comrade in the command module orbiting the Moon, the top section of the lunar module (**ascent stage**) blasted into space using the bottom section of the lunar module (**descent stage**) as a launch pad (Figure 5.30).

Once back in lunar orbit, the service module's engine was fired and the three astronauts returned to the Earth. Upon reaching the Earth, the command module detached from the service module and fell by parachute into the sea where they were picked up by helicopters and deposited onto an aircraft carrier.

The Results from Apollo; The Apollo missions have yielded a wealth of information concerning the Moon. The surface of the Moon, as it turns out, is covered by a layer of powdery soil which is between 10 and 100 feet deep. This **regolith** appears to be debris from the bombardment which formed the lunar craters. As already stated, the surface of the Moon consists of two types of regions: light colored, heavily cratered regions called **highlands** and dark, flat **lunar seas** or **lowlands**. The lunar highlands are composed, for the most part, of a low density rock called **anorthosite**. The ages of these rocks are up to four and a half billion years and hence are the oldest material on the Moon. It is hypothesized that at the time of the lunar origin, compression and massive impacts heated the Moon to the molten state. The lowest density rock crystals (called **feldspars**) floated to the surface and cooled to form the anorthositic lunar highlands. Further bombardment and heating caused fragmenting and cementing of rock particles into a rock type known as **breccias** (some breccias are also found on the lunar seas).

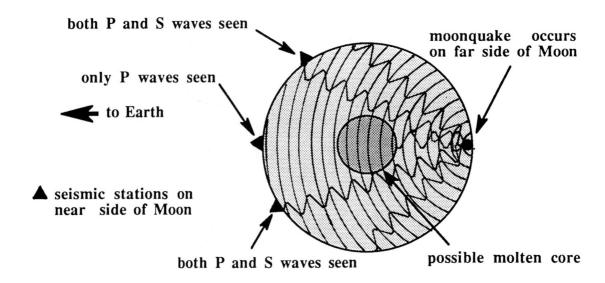

Figure 5.31: Seismic Studies of the Lunar Interior.

The lunar seas were formed when lava filled vast lowland areas from 3.1 to 3.9 billion years ago. This lava solidified into an igneous rock type known as **basalt**.

The lunar rocks themselves are similar to Earth rocks with certain exceptions. Low melting point materials ("volatiles") are strongly depleted or totally absent in Moon rocks (for example, water is totally absent) while high melting point materials (refractory elements, such as calcium, aluminum and titanium) are in relatively high abundance, except that iron is extremely depleted on the Moon. The implication of this is that the lunar material was intensely heated before the Moon formed. This heating would have driven the volatile elements into space.

The Lunar Interior; By placing seismographs on the lunar surface, scientists have been able to measure the ways in which shock waves transmit through the lunar interior. This, in turn, gives us information about the structure of the interior.

If a moonquake or an impact occurs on the far side of the Moon, the waves from this event will propagate through the Moon to the seismic stations on the near side. There are, however, two types of seismic waves: **P** (primary) waves which travel through both solids and liquids and **S** (secondary) waves which travel only through solids. Indications are that S waves cannot travel through the lunar core or inner mantle (**asthenosphere**) and this implies that this region is molten or semi-molten (Figure 5.31). Contradictory evidence, however, comes from lunar magnetic studies which indicate that the Moon has virtually no magnetic field. This fact argues against a large molten core as rotating molten cores are thought to generate such fields. It is possible that the molten core exists but that the slow lunar rotation is not enough to create the field.*

The mean density of the Moon is 3.3 grams/cc compared to the 5.5 grams/cc for the Earth. This indicates that the Moon is largely rocky, similar in density to the Earth's mantle. On Earth, quakes occur at the boundary between the lithosphere and the asthenosphere; the zone at which the Earth's crustal slippage takes place (i.e., continental drift). Since, on the Moon, quakes occur at a depth of about 600 miles, it is believed that this depth marks the bottom of the lithosphere although no continental drift is evident on the Moon (Figure 5.32).

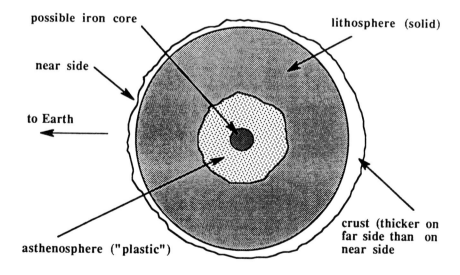

Figure 5.32: Hypothetical Cross Section of Moon

---

*It should also be noted that lunar surface rocks have a residual field far stronger than the Moon's current field. This could mean that when the rocks formed, billions of years ago, the lunar field was much stronger than it is now and that there was a molten core and a more rapid lunar rotation.

The Far Side of the Moon; When the lunar crust is studied, it is found that its thickness is less on the near side (about 40 miles) than of the far side (about 60 miles). This has caused a difference in appearance between the two hemispheres. Because of the thicker crust on the far side, lava has not been able to flow up through fissures in the crust to fill impact basins and form large maria. Compared to the near side, the far side of the Moon shows far more craters (Figure 5.33).

Figure 5.33: The Far Side of the Moon

The Origin of the Moon; Until recently, there were three theories of lunar origin. These theories were hotly debated and each had its proponents. Briefly, these three theories are:

- **The Fission Theory**: In this theory, the Moon was "spun-off" from the Earth at the time when the Earth was still in a semi-molten and rapidly rotating state (Figure 5.34).

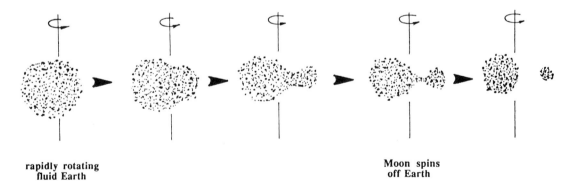

Figure 5.34: The Fission Theory

- **The Capture Theory**: The Moon formed at some other place in the solar system, wandered close to the Earth and was captured by the Earth's gravity (Figure 5.35).

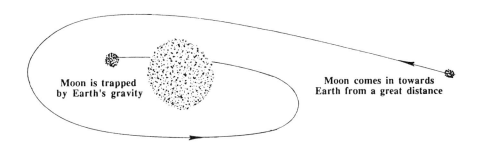

Figure 5.35: The Capture Theory

- **The Accretion Theory**: The Moon formed out of material which was left in orbit about the Earth after the formation of the Earth (Figure 5.36).

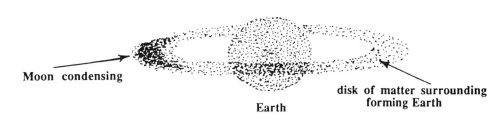

Figure 5.36: The Accretion Theory

Each of the above theories, as it turns out, cannot explain particular observations which have been made concerning the Earth and Moon. For the fission theory to be correct, the Earth would had to have been spinning with a period of 2 1/2 hours; 10 times more rapidly than at present. Theory indicates that the Earth was never spinning that fast. Also, if the Moon were "flung" off a spinning Earth, we might expect it to be orbiting the Earth's equator rather than being tipped to that equator by 18° to 28°.

The capture theory is unlikely because a body, falling towards the Earth from a great distance, would pick up too much kinetic energy to be captured into a closed orbit. Rather, unless it could lose some energy, it would fly off again into space.

The accretion theory cannot explain why the Moon, if it formed so near the Earth, has such a relatively low iron abundance.

- **The Impact Theory:** In the last 15 years, a 4th theory of lunar origin has gained wide acceptance. It is known as the **impact** hypothesis. It suggests that after the Earth had formed, and separated into an iron core and rocky mantle, it was struck in the upper mantle by a Mars-sized protoplanet. The impact blasted out hot rocky material from the Earth and the impacting body (Figure 5.37). Some of this material fell back to Earth, some escaped into space and some fell into an Earth orbit where it later condensed into the Moon. This theory satisfies our observations better than any prior theory. It explains why the Moon's mean density is so low and is depleted in iron (it formed from mantle material), why it is depleted in volatiles such as water (it formed from material which was heated by collision) and why its orbit lies closer to the ecliptic plane (5°) than to the equatorial plane (the body which struck the Earth came in from along the plane of the ecliptic). While the question is not settled, we are closer to the answer than ever before.

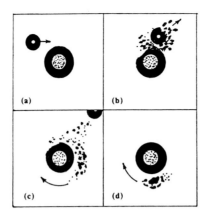

Figure 5.37: The Impact Theory

ECLIPSES

Shadows; Eclipses have to do with shadows, so before we begin with this topic it is best to get a few basic points straight. If a source of light is a point source, a body in the light path will cast a diverging dark shadow called the **umbra**. From any point within the umbra, the source will be eclipsed (covered) (Figure 5.38).

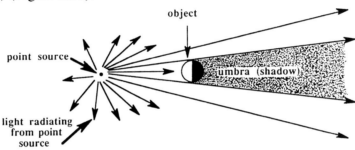

Figure 5.38: The Shadow from a Point Source

If, however, the source of light is an extended source, the shadow cast by the body will have three distinct regions. First, there will be a converging, dark, conical shadow again called the **umbra**. Within the umbra, no light from the source is seen. On both sides of the umbra there is a semi-dark, diverging region called the **penumbra**. Within the penumbra there is partial shadow; some, but not all, of the light from the source is seen (Figure 5.39). An extension of the lines drawn to form the umbra leads to the **region of transit**. Within this region, the body does not appear large enough to cover the source and hence the source appears as a ring (or **annulus**).

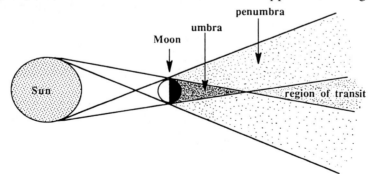

Figure 5.39: The Shadow From an Extended Source

Solar Eclipses; Solar eclipses occur when the Moon passes between the Earth and the Sun. This situation, which can only occur at times of new moon, is shown in Figure 5.40.

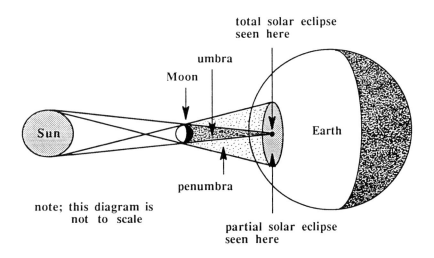

Figure 5.40: Circumstance of the Solar Eclipse

Notice that the umbra of the Moon's shadow just manages to reach the Earth. This is illustrative of the fact that, from the Earth, both the Sun and the Moon appear to be the same angular size in the sky. If one happens to be standing at that place on the Earth where the Moon's umbra touches down, one will see the Moon totally covering the Sun. On all sides of the umbra, the penumbra of the Moon's shadow touches the Earth. If one is standing in the Moon's penumbra, one sees the Moon partially covering the Sun. The two cases described above are called, respectively, **total solar eclipses** and **partial solar eclipses** (Figures 5.41 and 5.42). Whether one will see a total solar eclipse, a partial solar eclipse or no eclipse depends on where on the surface of the Earth you are located when the eclipse occurs.

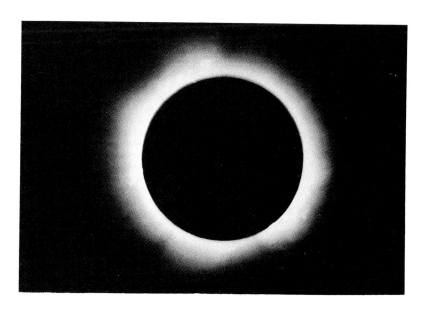

Figure 5.41: Total Solar Eclipse, March 7, 1970

A third type of solar eclipse is called the **annular solar eclipse**. This type of eclipse occurs when the Earth and the Moon are separated by too great a distance for the Moon's umbra to reach the Earth. This situation can occur since the Moon's orbit is elliptical and, hence, the Moon is sometimes farther from the Earth than at other times. If the eclipse occurs when the Moon is at or near its most distant point from the Earth (apogee), the Earth will find itself within the region of transit. In this case, the Moon does not appear large enough to cover the Sun completely and a ring of sunlight appears around the darkened lunar disk (Figure 5.43).

Figure 5.42: Partial Solar Eclipse

Figure 5.43: An Annular Solar Eclipse

Of the three types of solar eclipses, the total eclipse is the most beautiful and has the most scientific value. To observe this type of eclipse one must be at the right place at the right time. Figure 5.44 shows the total solar eclipse of June 8, 1918.

Figure 5.44: Total Solar Eclipse, June 8, 1918

The Path of Totality; The speed of the Moon in its orbit around the Earth is about 2000 miles/hour (eastward). The shadow of the Moon would, therefore, be expected to move across the Earth's surface at the same rate. The Earth, however, is rapidly rotating (also eastward). At the equator, the rate of rotation is just over 1,000 miles/hour. At the poles, the rate of rotation is zero. This causes the shadow of the Moon to move along the ground at a rate which depends on your latitude. At the equator, the rate is 1000 mi/hr (2,000 mi/hr minus 1,000 mi/hr). At the poles, the rate is 2,000 mi/hr (2,000 mi/hr minus 0 mi/hr) . The rate of speed of the shadow is between 1,000 and 2,000 mi/hr at intermediate latitudes.

If one wishes to see a total solar eclipse, one must station oneself somewhere on the eclipse path. These paths can be plotted in advance by astronomers who have a knowledge of the Moon's orbit about the Earth and the Earth's orbit about the Sun (Figure 5.45).

The duration of "totality" for a particular eclipse is determined by how large the umbral spot is and how fast the spot moves past your position. The diameter of the spot may be anywhere from 0 to 167 miles, depending on how far the Earth is from the Moon at the time of the eclipse (The penumbral spot generally extends about 2,000 miles on either side of the umbral path.). The velocity of the spot may be anywhere from 1,000 to 2,000 mi/hr depending on where on the Earth the spot touches down. Considering these two factors, totality may last for anywhere from zero to seven and a half minutes. A good average figure is about two to three minutes.

One may wonder, after all of this, about the value of total solar eclipses. To anyone who has seen one, a total solar eclipse needs no scientific justification; the sheer beauty of the event is beyond description. The wave of darkening pervading the landscape, the sudden drop in atmospheric temperature, the shadow-bands rippling across the ground, the emanations from the blue-white corona; these phenomena, and more, have consistently drawn thousands of observers and caused them to travel thousands of miles for no more than a few minutes of totality.

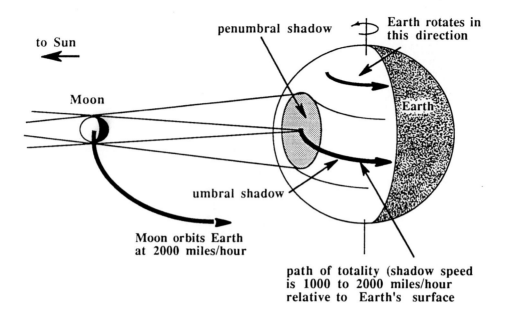

Figure 5.45: The Path of a Total Solar Eclipse

Aside from aesthetic considerations, total solar eclipses provide astronomers with a rare opportunity to observe the outer solar corona; that tenuous envelope of the Sun which reaches temperatures of 4 million degrees F. Only through these observations can scientists gather the data necessary to explain the conditions found in this part of the Sun.

Another experiment which has historically been performed during total solar eclipses is a photographic search for the **Einstein effect**. This effect relates to a prediction of General Relativity that the paths of light beams should be deviated by a near passage to the Sun. Scientists, hoping to verify this predication of the **Theory of Relativity**, try to detect a slight displacement in the positions of stars which can just be observed off the limb (edge) of the Sun during the total phase of the eclipse. The experiment was originally performed in 1919 by Sir Arthur Eddington and has been duplicated many times since. Today, confirmation of the Einstein effect is best accomplished by using radio telescopes.

Lunar Eclipses; Lunar eclipses occur when the Moon passes into the shadow cast by the Earth. This situation, which can occur only at full moon, is shown in Figure 5.46.

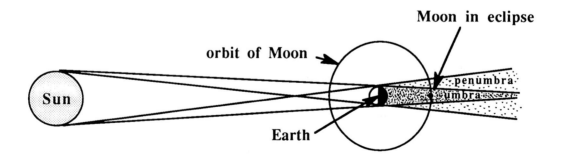

Figure 5.46: Circumstance of the Lunar Eclipse

As the Moon swings in its orbit about the Earth, it may, on certain occasions, cut through the penumbra and the umbra of the Earth's shadow (why this does not occur every month will be discussed in the next section). Figure 5.47 shows a cross-section of the Earth's shadow at the distance of the Moon.

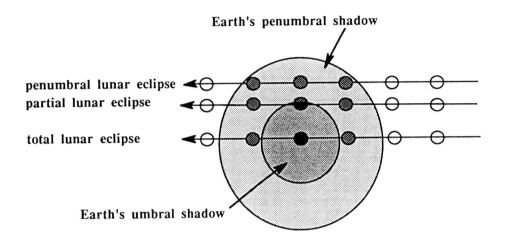

Figure 5.47: Cross-section of the Earth's Shadow

Path A represents the situation where the Moon cuts through the Earth's penumbra but not its umbra. This is called a **penumbral lunar eclipse**. In this case, the darkening of the Moon is so slight that it is hardly noticeable. This type of eclipse is generally ignored. Path B represents the situation where part of the Moon, but not all of the Moon, enters the umbra. This is called a **partial lunar eclipse**. Path C represents a **total lunar eclipse**. Partial and total lunar eclipses are known as **umbral lunar eclipses**. Contrary to what you might expect, during a total lunar eclipse the Moon turns a coppery red color. This is due to the fact that red light from the Sun is refracted through the Earth's atmosphere and strikes the Moon. Short wavelength (blue) solar radiation is "scattered" by the Earth's atmosphere and, hence, does not make it to the Moon. The blue scattered light is what gives the sky its blue color (Figure 5.48).

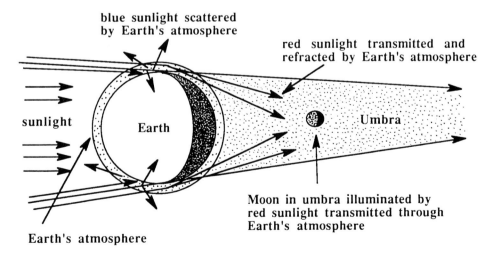

Figure 5.48: Illumination of the Moon During a Total Lunar Eclipse

Frequency of Eclipses; Why do we not have solar eclipses at every new moon and lunar eclipses at every full moon? The answer is that we would *if* the orbit of the Moon lay along the ecliptic (plane of the Earth's orbit). Actually, the Moon's orbit is inclined by 5° to the ecliptic so that the Moon is not necessarily directly between the Earth and Sun at new moon nor directly on the Earth-Sun line at full moon. If, at times of new moon, the Moon is on or near the ecliptic, an eclipse will occur. Another way of saying this is that if the **line of nodes** is pointing towards the Sun at new or full moon, an eclipse will occur.

There are two times during the year when the line of nodes points approximately towards the Sun. These two times are called **eclipse seasons**. For solar eclipses, in general, the eclipse season will last from 30 to 36 days. For total solar eclipses, the season will last for 20 to 24 days. Since a New moon must occur every 29 1/2 days, there must be at least one, and possibly two solar eclipses per eclipse season. The eclipse, however, may or may not be a total solar eclipse. The eclipse season for partial and total lunar eclipses is 19 to 24 days long. A lunar eclipse, therefore, may or may not occur during an eclipse season (Figure 5.49).

In any one calendar year there must be two to five solar eclipses (zero to three of the total variety) and zero to three umbral lunar eclipses. The reason that there may be five solar eclipses per year is that due to the regression of the line of nodes of the Moon's orbit, eclipse seasons occur about 20 days earlier each year. An eclipse season which occurs in early January of a given year will occur again in December of the same year (after the one that occurs in June).

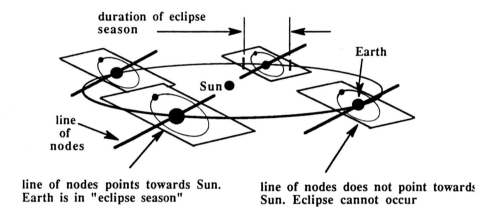

Figure 5.49: Eclipse Seasons

Eclipse Cycles; Because certain conditions must be satisfied before an eclipse can occur, we can predict future eclipses by understanding when those conditions will be met. The first requirement for a solar eclipse is a New moon which occurs every 29 1/2 days (the synodic month). The second requirement is that the Moon be at or near one of the nodes of its orbit. The period of revolution of the Moon with respect to the nodal points is 27.21 days and is called a **draconistic month.*** A solar or lunar eclipse will occur any time that these two cycles coincide. It turns out that 223 synodic months equals 6,585.321 days while 242 draconistic months equals 6,585.781 days. The two cycles, therefore, come into phase (almost) every 18 years, 11 days and, at that time, an eclipse will take place. This time interval is known as the **Saros**. The July 11, 1991 total solar eclipse was of the same Saros cycle as the Great Sahara Desert eclipse of June 30, 1973 and as the 1919 eclipse when Arthur Eddington confirmed the Einstein effect. The next eclipse of the series will be on July 22, 2009. There are several Saros cycles operating at the same time and a total solar eclipse will occur on Earth about every 1 1/2 years. Figure 5.50 shows future total and annular solar eclipses.

*The draconistic month is less than the sidereal month due to the regression of the line of nodes of the Moon's orbit.

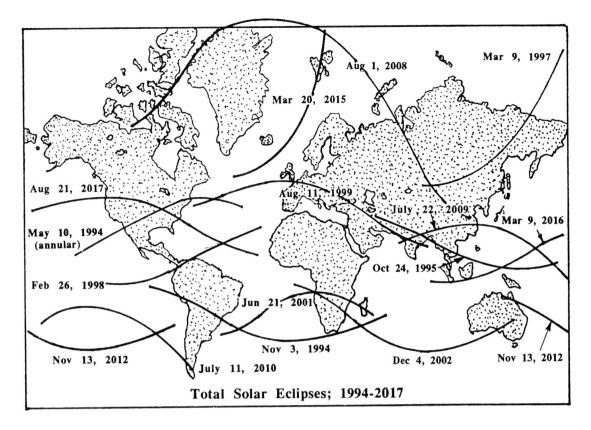

Figure 5.50: Paths of Future Solar Eclipses

Visiblity of Eclipses; Even though total and annular solar eclipses (collectively called **central eclipses**) occur with about the same frequency as partial eclipses, they are not equally visible. The path of a central eclipse is relatively narrow, only a few thousand miles long and, thus, an observer at a particular location on Earth has only a small chance of seeing that eclipse. Statistically, a particular location on Earth will experience a total solar eclipse only once in 400 years. By contrast, a partial solar eclipse is seen every two years or so and a lunar eclipse about once a year.

Moreover, since during a lunar eclipse more than half of the Earth can see the eclipsed Moon, Earth-bound observers are far more likely to see a lunar eclipse than a solar eclipse.

TIDES

Tides are caused by the *difference* between the gravitational forces on the opposite sides of a body. As shown in Figure 5.51, the force of attraction between mass M and body 1 is greater than the force of attraction between mass M and body 2. From the point of view of the center of mass, these two forces tend to separate bodies 1 and 2.

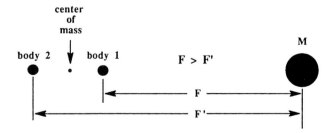

Figure 5.51: Tidal Force

If M is the Moon and bodies 1 and 2 are two points on opposite sides of the Earth, the effect of the gravity of the Moon is to "stretch" the Earth. This stretching affects the land, air, and the water of the Earth (Figure 5.52). In a similar way, the Earth raises tides on the Moon. The Sun also raises tides on the Earth, but because of its great distance, the differential force is small and the solar tides end up being about half as strong as lunar tides. When the Sun and the Moon are pulling along the same line (at new and full moon) the tides felt by the Earth are maximized. This situation is called **spring tide**. At quarter moon, the Moon and Sun are pulling at right angles to one another and the tides are minimized. This is called **neap tide** (Fig 5.53).

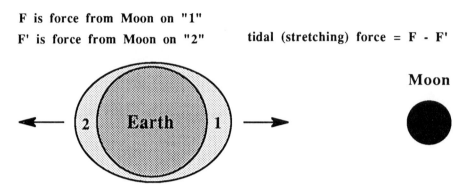

Figure 5.52: Tidal "Stretching" of Earth

Because the day and the month are of unequal duration, the Earth's tidal axis is not along the Earth-Moon line of centers (Figure 5.54). The reason for that is that while the Earth's water tries to flow to the point which is directly under the Moon (and to the point which is 180° from the sub-lunar point), the rapid turning of the Earth tries to take the water away from this point. Naturally, the Earth exerts dominant control over the water and the bulge can never "reach" the line of centers.

In its effort to "get under the Moon", the waters of the Earth move across the ocean basins (i.e., the tides come in and out). The friction between the sea floors and the water is causing the Earth's period of rotation to increase by about two milliseconds per century. This slowing down of the rotation of the Earth has been verified by several independent means.

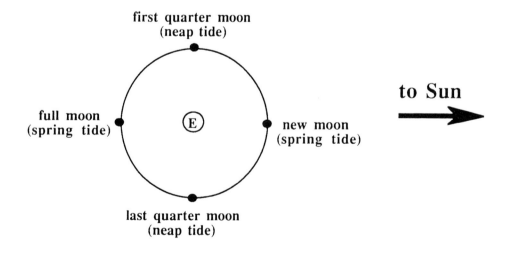

Figure 5.53: Spring and Neap Tides

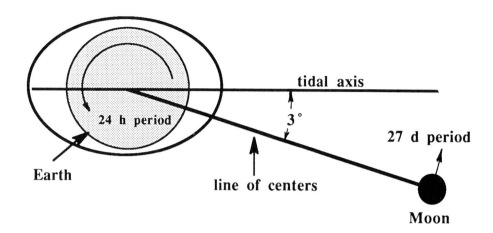

Figure 5.54: Tidal "Lag" of Oceans

The principle of conservation of angular momentum dictates that the Moon cannot rob the Earth of its spin energy without being affected itself. The theoretical result of this tidal interaction is that while the Moon slows the rotation of the Earth, the Moon itself will speed up in its orbit. This will cause the Moon to spiral away from the Earth and cause the length of the month to increase. The rate at which the Moon is receding from the Earth is predicted to be of the order of 1 inch per year. This has been confirmed by lunar laser ranging.

While both the length of the day and of the month are increasing, the day is increasing more rapidly and it is calculated that in the future, the day will "catch" the month at a value of 40 days. When this occurs, the same side of the Earth will always face the Moon (synchronous rotation) and the lunar tides will lose their effectiveness at slowing the Earth further. The Sun, however, will still be a tide producing agent and will further cause the Earth to slow. This will cause the day to be longer than the month which in turn will cause the axis of the Earth's tidal bulge to lag behind rather than lead the Earth-Moon line of centers. When this occurs, the effect of the lunar tidal friction will be to speed the Earth's rate of rotation, decreasing the length of the day. In response to this, the Moon will be slowed in its orbit and forced to spiral in towards the Earth. The ultimate destiny of the Moon may be to get within **Roche's limit** of the Earth, the distance at which gravitational tidal forces will disrupt a large body held together solely by gravitational forces. Roche's limit is given by the expression $L_R = 2.8r$, where r is the radius of the tide producing body. Since for the Earth, $r = 4,000$ miles, L_R will be about 11,000 miles. If the Moon approaches to within this distance, it will be ripped apart by tidal forces and form a ring system around Earth similar to Saturn's. If this hypothetical scenario is correct, the time scale for the above tidal evolution is tens of billions of years.*

---

*It should be noted that long before this occurs, the sun will die and, in doing so, may vaporize both the Earth and the Moon.

CHAPTER 6

THE SUN

The dominant body in the Solar System, containing 99.8% of the Solar System's mass, is the Sun. It is the gravity of the Sun which keeps all of the bodies discussed in previous chapters, from drifting off into interstellar space. The radiation (light) from the Sun is responsible for warming the inner solar system and, therefore, maintaining life on Earth. The Sun has been presiding over the solar system for about five billion years and modern estimates indicated that it will continue to do so for another four or five billion years.

Yet despite its apparent singular nature, the Sun is just one of billions of similar bodies which populate our galaxy. The Sun is a star. It is slightly larger and brighter than average, yet it is just a typical star like those seen dotting the night sky. It appears different from the stars we see at night only because of its proximity to Earth.

The study of the Sun is an important task for the astronomer. Not only is it important to understand the nature of the body on which we depend for life, but the Sun gives us a chance to study a star at close range and, by doing so, better understand the bodies which are the building blocks of the galaxies and hence the universe (Figure 6.1).

How We Study the Sun; In some ways, studying the Sun is more difficult than studying the planets and satellites and, in some ways, it is easier. Clearly, we cannot land on the Sun the way we landed on the Moon, Mars and Venus. A spacecraft would be vaporized long before it could closely approach this stellar furnace. It is, therefore, not possible to immerse ourselves in the layers of luminous solar surface which we call the **photosphere** (Figure 6.1). Yet the Sun has an outer atmosphere which we cannot see with the naked eye. It is called the **solar wind**, and the entire Earth is immersed in this tenuous outer envelope of the Sun. Studies of the solar wind, generally carried out by Earth satellites and planetary probes, give us first hand information concerning the nature of the outer layers of the Sun.

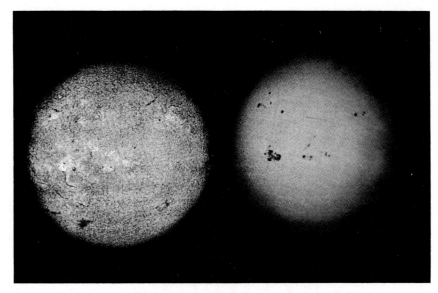

Figure 6.1: The Sun; August 12, 1917
(R) Visual
(L) Hydrogen Light (Hα)

Furthermore, there are many bodies in the solar system that we know much about yet on which we have not landed. This is the case since the light which reaches us from these objects carries much information about the nature of the object. This is particularly true if we study the "emitted" radiation from the body.

In general, bodies will "reflect" radiation from their surfaces and, if the bodies are warm, "emit" radiation from their surfaces. An example of reflected light is the light you see coming from the Moon or from the shirt of the person next to you. An example of emitted light is the light you see coming from a light bulb or from the Sun. Radiation "emitted" by a body tells us much about the nature of the body. This is true because as certain characteristics of the body change, so will the nature of the emitted radiation. To understand this principle, we must first understand something about the nature of light.

The Electromagnetic Spectrum; As Isaac Newton revealed in the 17th Century, sunlight is a mixture of colors known as the **visible spectrum**. It was later discovered that "colors" beyond the visible exist. These other types of light (technically known as electromagnetic radiation) include **infrared, microwaves** and **radio waves** (on the "red" side of the visible spectrum) and **ultraviolet, x-rays** and **gamma rays** (on the "blue" side of the visible spectrum). A schematic diagram of the full electromagnetic spectrum is shown in Figure 6.2.

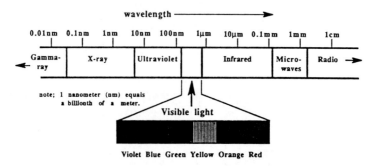

Figure 6.2: The Electromagnetic Spectrum

Although the exact nature of electromagnetic radiation is rather complex, it is possible to describe many of its general properties in terms of what is known as the **wave model**. In this concept, developed in the 19th century, electromagnetic radiation consists of a series of electrical and magnetic vibrations which propagate through space at 186,000 miles/second ("the speed of light"). A schematic of one of these vibrations (waves) is shown in Figure 6.3.

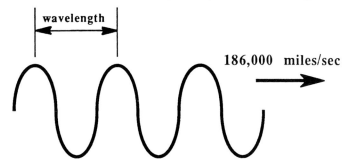

Figure 6.3: An Electromagnetic Wave

The distance between successive "crests" or high points in the wave is called **wavelength**. According to the wave model, the different colors of the visible spectrum as well as the various invisible electromagnetic radiations (e.g. radio, x-ray, etc.) are simply waves of different length. The human eye is sensitive to the wavelengths between 0.000016 inch ("violet" light) and 0.000028 inch ("red" light) and cannot see either longer or shorter wavelengths.

Spectroscopy; It is now known that bodies at any temperature emit electromagnetic radiation. The quantity and quality of this emission depends on the temperature of the emitting body. This idea is expressed in Figure 6.4; a graph known as **Planck's law**.

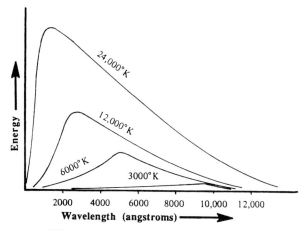

Figure 6.4: Planck's Law

It is seen from Planck's law that bodies of any temperature emit electromagnetic radiation (**EMR**), but that hotter bodies emit more EMR than do cooler bodies. Notice also that hot bodies do not emit uniformly at all wavelengths but that there is a particular wavelength at which the output "peaks" (e.g. yellow for a 10,000° F body). It is this property of hot bodies which causes them to change color as they are heated. Bodies at 6,000° F appear red, while bodies at 40,000° F appear blue. It is, therefore, possible to determine the temperature of a distant hot body (the Sun or another star) by determining the color of light which predominates in its spectrum. Since the predominant color in the solar spectrum is yellow, the Sun must be about 10,000° F.

The light spectrum of the Sun can also be used to determine its chemical composition. It has been found that low pressure gases, like those in the solar photosphere (atmosphere), are capable of absorbing particular colors of light and that the particular colors which are absorbed are determined by the gas doing the absorbing. In this way, hydrogen will absorb (from the visible spectrum) a certain wavelength of red light, a certain wavelength of blue-green light and a certain wavelength of violet light. Thus, the **dark line spectrum** of hydrogen (Figure 6.5) will be characteristic of that gas. All other gases absorb a set of colors which is characteristic of the particular gas.

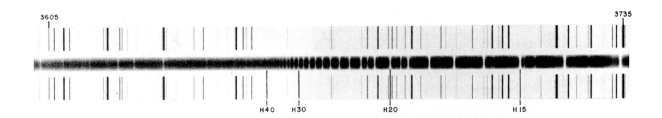

Figure 6.5: The Dark Line Spectrum of Hydrogen

When the light spectrum of the Sun is observed and recorded, it is found that the chemical composition of the atmosphere can be deduced from the particular **dark lines** (missing wavelengths) in the spectrum. The solar dark line spectrum is found in Figure 6.6.

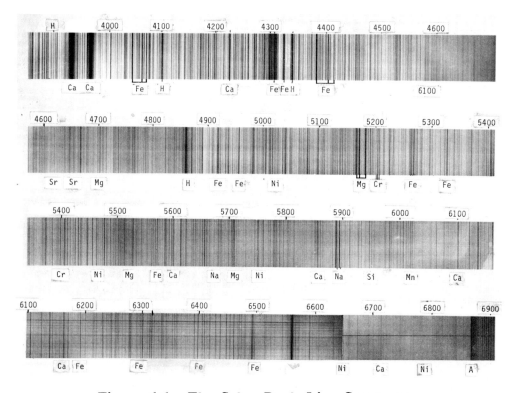

Figure 6.6: The Solar Dark Line Spectrum

In the case of the Sun, many elements are acting together to produce a composite spectrum. Therefore, there will be many hundreds of dark lines. It is the task of the astronomer to "untangle" the puzzle to determine just which elements, in what abundances, must have acted to produce the observed spectrum.

Observations of the radiation emitted from the Sun have revealed much about our star. The following sections of this chapter summarize what has been discovered about the various parts of the Sun.

Astronomers have found it useful to divide the sun into six sections because of the divergent properties within these regions. These sections are (from inside to outside); the **fusion core**, the **envelope**, the **photosphere**, the **chromosphere**, the **corona** and the **solar wind**. Of these sections, only the core and envelope cannot be directly observed and hence our knowledge of them must be founded on largely theoretical grounds. The fusion core and envelope will be discussed last. Figure 6.7 is a schematic diagram of the Sun and its regions.

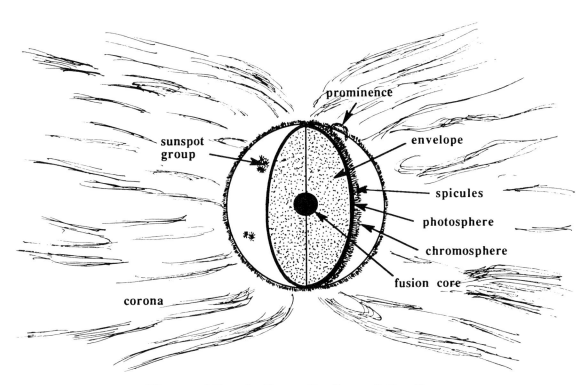

Figure 6.7: A Cross Section of the Sun

The photosphere is that part of the Sun which you see when you look at it with your unaided eye.* The chromosphere lies immediately above the photosphere and can only be seen with special instrumentation or during a total solar eclipse. The corona is the distended outer atmosphere of the Sun and is best observed during a total solar eclipse. The outer extent of the corona is known as the solar wind. This "wind" from the Sun extends at least as far as Saturn and probably as far as Pluto.

The Solar Photosphere; The photosphere is that part of the Sun which emits the bulk of the visible light received at Earth. It is a gaseous layer about 200 miles deep which, because of

---

*One must take great care when observing the Sun. The Sun should <u>never</u> be observed through binoculars or a telescope unless these observations are guided by an experienced astronomer with proper safety equipment.

its chemical makeup, is opaque to visible light. We are, therefore, unable to look through this layer to deeper regions in the solar interior. The temperature of this region may be ascertained by determining the wavelength at which the photospheric emanations peak (Figure 6.4). Such studies indicate that the top of the photosphere is at a temperature of about 9,000° F while 200 miles down, the temperature has climbed to over 13,000° F. The density of the photospheric gases is about one millionth the density of water and the gas pressure is several hundred thousand times less than sea level air pressure on Earth. The diameter of the photosphere is about 864,000 miles. Analysis of the solar dark line spectrum has revealed the composition of the gases in the photosphere. This composition is given in Figure 6.8.

Element	Percent of Sun (by mass)
Hydrogen (H)	76.4
Helium (He)	21.8
Oxygen (O)	0.8
Carbon (C)	0.4
Neon (Ne)	0.2
Iron (Fe)	0.1
Nitrogen (N)	0.1
Silicon (Si)	0.08
Magnesium (Mg)	0.07
Sulfur (S)	0.05

Figure 6.8: The Composition of the Sun

It is seen that the Sun is about 98.9 percent hydrogen and helium and that the remaining one percent is divided among over 60 other elements. Certain phenomena seen within the photosphere are worthy of note:

Photospheric Granulation; When observed with special equipment, the "surface" of the photosphere appears "mottled" or "granulated". The granules, which are 600 miles to 1,000 miles in diameter, are rising and falling columns of gas. Typical column velocities are one to two miles/second. The granules are caused by the changing temperature as we move downward through the photosphere. Since the gas at the bottom of the photosphere is hotter than the gas at the top, rising columns from below bring hotter, brighter gases to the photospheric surface. These are the light colored granules. The darker regions between the bright granules are cooler gases falling to the bottom of the photosphere replacing the hotter gases which have risen to the surface (Figure 6.9).

Sunspots; Sunspots are highly magnetic, relatively cool regions in the photosphere. While the photosphere itself averages 11,000° F, the spots are from 8,000° F to 9,000° F and, hence, appear darker than their surroundings. They live for anywhere from hours to months and sometimes grow as large as 100,000 miles in diameter (Figure 6.10).

A sunspot consists of two regions; a dark center (**umbra**) and a lighter gray perimeter (**penumbra**) (Figure 6.11). It has been found that near the surface of the photosphere, gases are flowing horizontally away from the spot while above the photosphere, gases are flowing horizontally towards the spot. These velocities are less than one mile/second. Bright regions surrounding the sunspot or sunspot group are called **plages**.

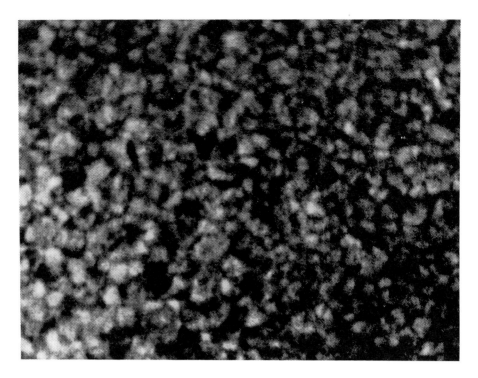

Figure 6.9: Photospheric Granulation

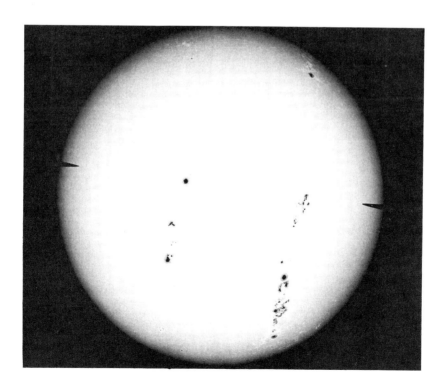

Figure 6.10: The Sun; February 20, 1956

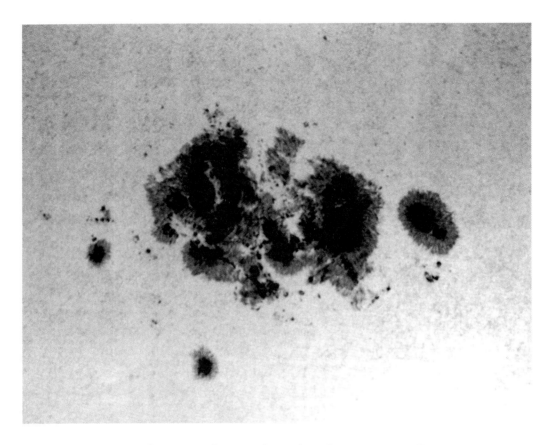

Figure 6.11: Sunspot Group Showing Umbra and Penumbra

Sunspots frequently occur in groups of two or more. Since the Sun is rotating, the spots in a sunspot pair are designated as the preceding (**P**) spot and the following (**F**) spot. Sunspots are highly magnetic and the P and F spots in a typical group are generally found to have opposite polarity (one spot will be a north magnetic pole and the other will be a south magnetic pole). The polarity of the "P" spots will all be the same as the general polarity of the pole in the hemisphere in which the spot group is found (Figure 6.12).

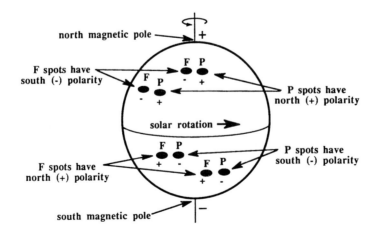

Figure 6.12: The Magnetic Polarity of Sunspot Groups

The Sunspot Cycle; In the middle of the 18th century, it was discovered by the German amateur astronomer Heinrich Schwabe that the number of sunspots visible in the solar photosphere varied with time; sometimes, many spots were visible while at other times, none were seen. Careful study showed that the period of this variation was between 8 and 18 years with an average of about 11 years. This variation is known as the **sunspot cycle** and is shown in Figure 6.13 (Learning Activity 13, Appendix IV).

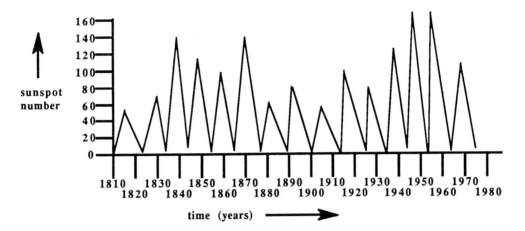

Figure 6.13: The Sunspot Cycle

At the beginning of a new cycle (just after sunspot minimum) a few spots appear at mid-solar latitudes (at about +30° and -30°). This is shown in Figure 6.14a. As the cycle progresses, new spots appear at lower and lower latitudes until the maximum of the cycle when spots appear near latitudes +/-15° (Figure 6.14b). Near sunspot minimum, the few spots which appear are found within 5° to 10° of the equator (Figure 6.14c).

At sunspot minimum, the overall general solar magnetic field fades and when the new cycle begins (with a few spots appearing at high latitudes), it is reversed in polarity (Figure 6.14d). Also notice in Figure 6.14d that the polarities of the P and F spots have reversed relative to the preceding cycle.

The sunspot cycle may also be viewed pictorially by the use of a diagram developed by **E. W. Maunder** in 1904; the **butterfly diagram** (Figure 6.15). In this diagram, the horizontal line represents the solar equator. The horizontal axis is a time axis and the vertical axis represents solar latitude. Each dot is a sunspot plotted as a function of its location on the Sun and the time of its appearance.

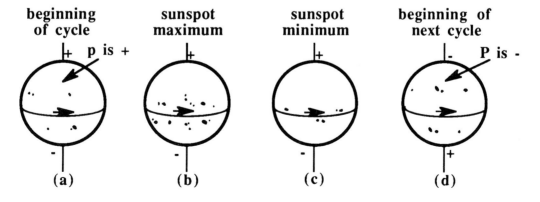

Figure 6.14: Sunspot Cycle

141

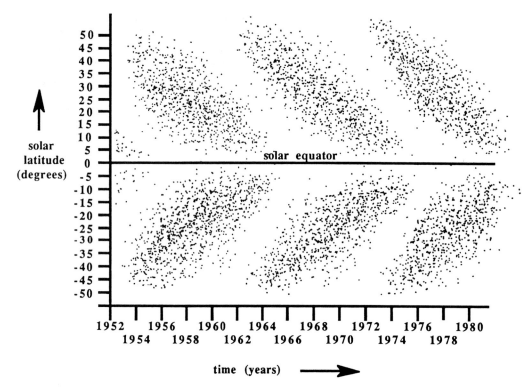

Figure 6.15: The Butterfly Diagram

Astronomers differentiate between the period of the sunspot cycle (time from sunspot minimum to sunspot minimum) which averages 11 years and the solar magnetic cycle (the time for the magnetic polarities to go through one complete cycle) which averages about 22 years.

In recent years it has become evident that the sunspot cycle is itself cyclic. Some maxima are higher than others. This can be seen in Figure 6.16 which shows that there is a long period variation in the "peaks" of the sunspot numbers. The period of this variation is about 80 years. In fact, records show that from about 1645 to 1715, the sunspot activity was essentially zero. This so-called **Maunder minimum** coincides with a period of extreme cold in Europe (the **little ice age**) and may hint at the connections between solar activity and terrestrial weather patterns.

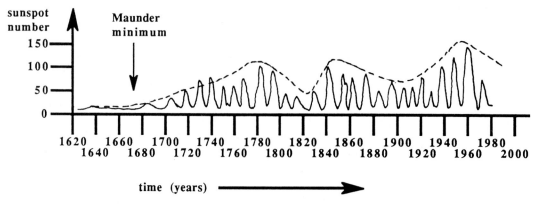

Figure 6.16: Variations in the Sunspot Cycle

While the cause of sunspots and the sunspot cycle is not known with certainty, an interesting theory for the mechanism was suggested by Horace Babcock in the early 1960's. This theory can be best illustrated by showing a series of sketches of the Sun; each sketch being separated by about one year of time (Figure 6.17). To understand the theory, one must first understand two things about the Sun. First, the Sun does not rotate as a solid body. The Earth, for example, is essentially solid (at least its crust is) and thus all parts have the same rotation period; it takes 24 hours for both Miami and Fairbanks to make one revolution about the Earth's axis.

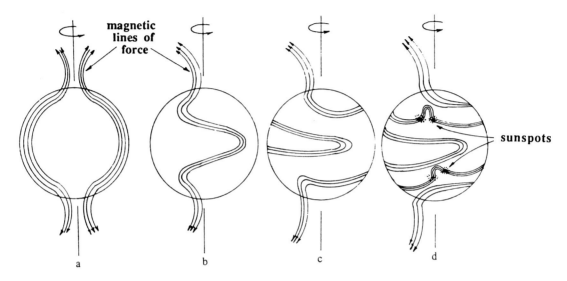

Figure 6.17: The Babcock Theory of the Sunspot Cycle

The Sun, however, is a gaseous body and rotates in a different manner. The equatorial regions of the Sun have a 25 day rotation period while at a latitude of 40 degrees, the period increases to 28 days. At higher latitudes, the period is even longer. This type of **differential rotation** is characteristic of gaseous bodies like the Sun and Jupiter.

The second important factor is that the Sun has a magnetic field which may be represented schematically by **lines of force**. These lines are shown in Figure 6.17. It is theorized that the Sun's magnetic field (which is equivalent to the material within the Sun which is magnetized) travels not through the center of the Sun but, rather, just under the surface.

As the Sun undergoes its differential rotation, the lines of force near the equator will "get ahead" of the lines of force at higher latitudes (Figure 6.17b). This is true because the equator rotates faster than other latitudes and the magnetic lines (magnetism) are trapped in the solar subsurface material.

As more and more time goes by, the equatorial lines will gain more and more on the higher latitude lines. After about a year, the equatorial lines will "lap" the higher latitude lines (Figure 6.17c). After several years, this lapping will take place several times and this will create a concentration of the magnetic lines. This concentration will occur first at about latitude 40 degrees. With sufficient magnetic concentration, a sort of buoyancy occurs which carries the magnetic lines and their included material to the surface of the Sun. The place where the lines "breach" the solar surface is the location of the magnetic disturbance which we call a sunspot (Figure 6.17d). As the winding of the lines of force continues, the critical magnetic concentration is reached at successively lower and lower latitudes, causing spots to appear at these latitudes. Toward the end of the cycle, the magnetic concentrations, and hence the spots, are appearing rather close to the equator.

Many other properties of the sunspots and of the solar magnetic field may be explained by this theory. Other properties of the sunspots and sunspot cycle, however, cannot be explained.

The Solar Chromosphere; Above the photosphere is a region which is generally transparent to the photospheric radiation; hence it is not easily viewed without special equipment. The **chromosphere**, which extends some 1,000 to 2,000 miles above the photosphere, is a thin, low density medium whose temperature increases with distance from the photosphere (from 9000°F to over 30,000° F). The chromosphere emits so much less light than the photosphere that it may only be seen when the disk of the Moon blocks the photosphere (**total solar eclipse**) or by using a special telescope with an artificial photospheric occulting device (a **coronagraph**). The spectrum of the chromosphere is made up of bright lines characteristic of specific elements. In 1868, the element **helium** was discovered in the chromosphere before its discovery on Earth.

Extending from the top of the photosphere, up through the chromosphere, we find many "jets" or "columns" of gas. These columns are called **spicules**. Spicules shoot out into the corona at about 20 miles per second and attain heights of up to 5,000 miles before they fade away. The diameter of a spicule is about 500 miles. The spicules are cooler than the surrounding coronal medium (Figure 6.18).

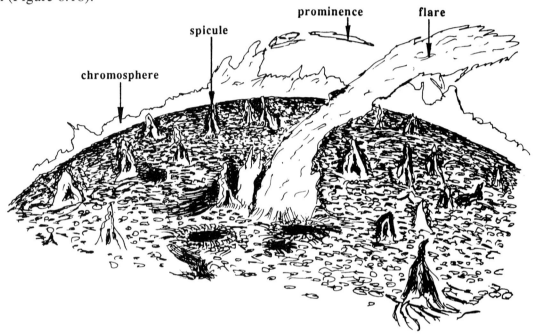

Figure 6.18: The Solar Chromosphere

Occasionally, the emissions from the solar chromosphere in a small region brighten to a high intensity. Such an event is called a **solar flare**. Charged particles as well as high energy electromagnetic radiation are emitted during such flares and some of these radiations and particles may reach the Earth, causing radio blackouts and auroral displays. Flares are generally associated with regions of sunspot activity and are, therefore, correlated with the solar cycle (Figure 6.19). Flares can last for minutes to hours before they fade away.

The Solar Corona and the Solar Wind; Extending outward from the chromosphere is the **corona**, the outermost envelope of the Sun (Figure 6.20). While the corona is rather static near the Sun, at greater distances its internal pressures are greater than the Sun's gravity and the corona starts expanding. The farther we go from the Sun, the faster is the rate of coronal expansion and the lower is the coronal density. The expanding part of the corona (beyond 2 or 3 million miles from the solar surface) has been given the name **solar wind**.

The composition of the solar wind is primarily protons and electrons. Near the Earth, the speed of the wind averages about 300 miles/second but "gusts" of up to 500 miles/second have been measured.

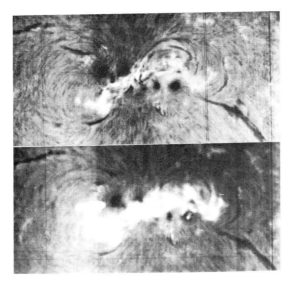

Figure 6.19: Sunspot and Flare, July 25, 1946

The wind has a strong effect on the bodies of the solar system. The gas or ion tails of comets are formed by the interaction of the solar wind with the comet (Figure 6.21). During strong solar flare bursts, the solar wind enhances and causes magnetic storms in the Earth's upper atmosphere, auroral displays and radio communication blackouts. The magnetic field carried outward into space by the solar wind to some extent shields the solar system from cosmic rays originating in outer space.

The corona and the solar wind are essentially a plasma cloud; an atmosphere of charged particles (electrons and atomic nuclei). As one travels upward through the corona, the temperature of the gaseous medium increases, reaching a temperature of about 4 million° F some 150,000 miles above the photospheric surface. The cause of this high temperature is unknown, but its effect is that the corona gas emits mostly x-rays. X-ray photos of the sun, taken from rocket or satellite, reveal that the corona is of uneven temperature. Near the solar poles, the corona is relatively cool and forms what is called **coronal holes** (Figure 6.22). These "holes" are just cooler parts of the corona. The coronal structure is maintained by the solar magnetic field. **coronal streamers** (Figure 5.44) highlight this magnetic field.

Like the chromosphere, the corona may only be seen by special instrumentation, or during a total solar eclipse.

The most spectacular coronal feature are **prominences**; relatively cool regions which are capturing electrons and emitting light. Prominences appear to be red, flamelike protuberances arching away from the Sun. These arcs of light-emitting gas have been seen racing outward into the corona at speeds of up to 1,300 miles per second (eruptive prominences). Some prominences have reached heights of one million miles above the solar photosphere. Prominences usually originate near regions of high sunspot activity and may be supported by the magnetic fields associated with these regions (Figures 6.24 and 6.25). Some prominences are static (quiescent prominences), holding their same position for hours to weeks.

Figure 6.20: The Solar Corona

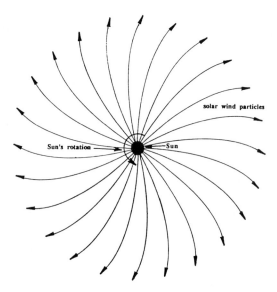

Figure 6.21: The Solar Wind

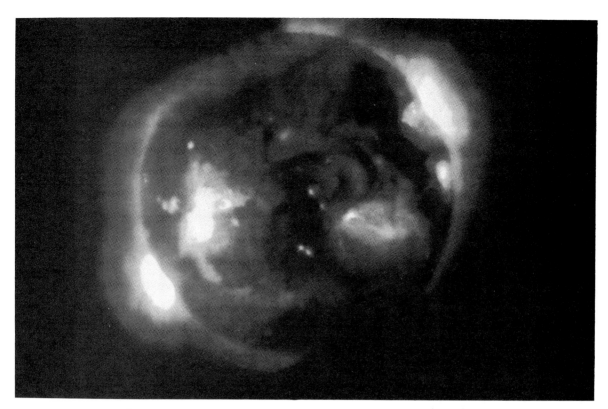

Figure 6.22: The X-ray Corona and Coronal Holes

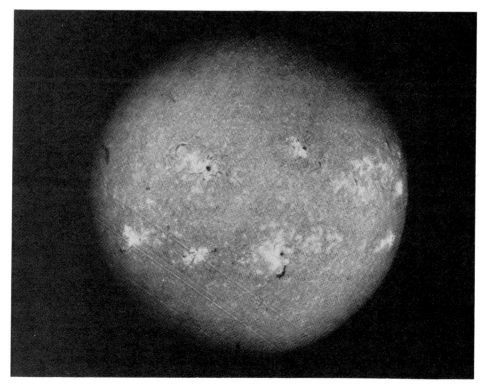

Figure 6.23: The Chromosphere, September 4, 1973

Sometimes, prominences are seen projected on the solar disk where they appear as black lines. These projections are called **filaments**.

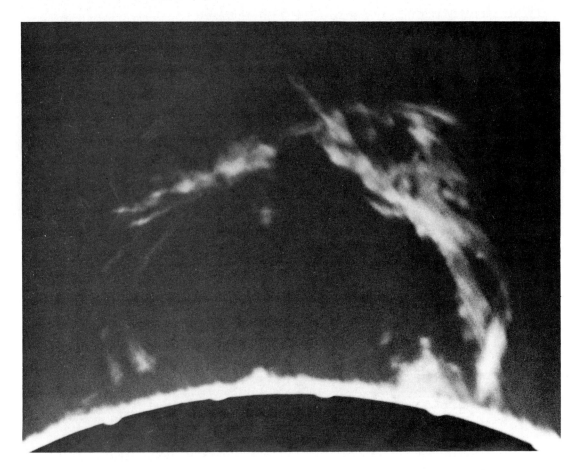

Figure 6.24:   Prominence above the Solar Chromosphere

The Solar Interior; Over the centuries, many theories have been developed concerning the source of the Sun's energy. It was not until the 1920's, however, that a concept came to light which explained all of the observed characteristics of the Sun. This theory, an outgrowth of Einstein's Theory of Relativity, suggests that the Sun is powered by nuclear reactions in the solar core. The particular nuclear reaction found in the Sun is known as **hydrogen fusion**.

Nuclear Fusion; A typical atom consists of a nucleus surrounded by a cloud of electrons. The nucleus contains protons and neutrons and the number of protons and neutrons in the nucleus determines the particular element (e.g. hydrogen, helium, carbon, gold, etc.). Because nuclei are positively charged (they contain protons), they tend to repel one another. But if they collide with great force, they can fuse together and become a single larger nucleus. This only happens at very high temperatures (millions of degrees F). When fusion occurs with four hydrogen nuclei, the element helium is formed and electromagnetic radiation is released. This fusion reaction, requiring 20 to 30 million degrees F, is apparently going on in the solar core. The energy released in the reaction makes its way to the solar surface where it is released as light and heat. This is the radiation which sustains life on Earth.

The fusion reaction rate inside the Sun is quite intense. Every second, 600 million tons of hydrogen fuel are consumed leading the solar power output of 400 trillion trillion watts.

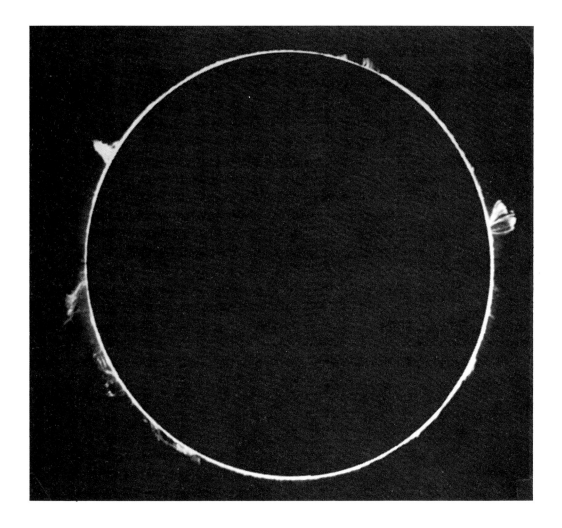

Figure 6.25: Prominences on the Solar Limb

The Future; Calculations have indicated that the Sun, when first formed, had enough core hydrogen for 10 billion years worth of fusion. Since the solar system appears to be about five billion years old (this is determined from the ages of Moon rocks, Earth rocks and meteorites), the Sun probably has about 5 billion years worth of fuel remaining.

When the hydrogen fuel is depleted in the solar core, the fusion reaction will die out and the core will gradually collapse. As it does so, the solar envelope (the region surrounding the core) will be pushed outward into space. The outer envelope of the Sun will grow larger and larger until the inner planets (including Earth) are incorporated into the solar atmosphere. The Earth will be no more and the Sun, after it pushes its envelope into space, will cool and die.

CHAPTER 7

THE ORIGIN OF THE SOLAR SYSTEM

Our knowledge of the origin of the Sun and its system of planets is necessarily limited by the fact that the event took place at a time deep in the past. Evidence indicates that our solar system was born about 4.6 billion years ago. No one was there to witness or to write about the happenings of that epoch and many changes have occurred in our planetary system since that time. Add to this the fact that distance makes it quite difficult to observe other stellar or planetary systems in the act of formation so that the **Solar System** is today the sole "data point" in planetary origin theory.

Despite these difficulties, **origin theories** have abounded over the past two centuries not only because we are curious about our beginnings, but because we are curious about the chances of finding other planetary systems, other "Earths", and other life forms. This chapter speaks to that curiosity.

Any theory of the origin of the Solar System must explain several "regularities" found in our planetary systems. The most important of these regularities are:

1. **The planetary orbits are generally coplaner (i.e. the Solar System is a flat place).**
2. **The planets all orbit the Sun in the same direction (counterclockwise when viewed from the north).**
3. **The planets orbit the solar equator.**
4. **The planets (except for 3) rotate in the same direction as they revolve. The Sun also rotates in that direction.**

5. The small, high density worlds are found nearest the Sun.
6. The large, low density worlds are found farthest from the Sun.
7. The planetary distances from the Sun obey Bode's law.
8. As we get farther from the Sun, the bodies we encounter contain larger percentages of water ice. At the farthest distances, comets are almost 100% water ice.
9. Comet orbits are spherically distributed about the plane of the Solar System.

If an "origin" theory cannot explain all or most of these Solar System's "facts", then it cannot be taken as a serious theory. After many decades of theorizing, observing, and checking, a basic scenario has developed. While it should not be taken as "certain", it explains many of the Solar System regularities so well that it must be taken as a serious contender for "the truth".

Before The Beginning; Modern astronomical observations reveal that the plane of our galaxy is studded with huge gas clouds which astronomers call **nebulae** (plural for nebula) (Figure 7.1). Within these nebulae, astronomers have observed local condensations of matter; globules of gas and dust which seem to be in the process of collapsing under the influence of gravity to form stars and star systems like our Solar System.

Figure 7.1: The Orion Nebula

If these observations are correct, they argue that star systems (the Solar System included) are formed by the gravitational collapse of gas clouds in space. Further, each gas cloud seems capable of giving birth to hundreds or thousands of star systems. The only question which plagued astronomers was, "How did the collapse get started"? One theory held that old, massive stars, "living" near nebulae and dying in explosive events called **supernovae**, might send waves of energy through these nebulae and compressing the gas clouds to the point where new star systems might be initiated. Could it be that the death of an old star might be the cause of the birth of hundreds of thousands of new stars? In recent years, astronomy has provided a potential confirmation.

It has long been understood that the radioactive elements in our universe are created in the catastrophic detonations of stars (supernovae). This is the only scientifically reasonable genesis of this material which is then expelled into space by the explosive death of the star. The circumstantial confirmation of the **supernova genesis theory** of star system formation came with the analysis of the famous **Allende meteorite** which fell to Earth in 1969. This type meteorite is primordial material; material unchanged by chemical processes during the last 4.6 billion years. When the meteorite was chemically analyzed, a curious excess of an isotope of magnesium was discovered. But this isotope of magnesium is most generally the result of the radioactive decay of aluminum 26; a material created during supernova outbursts. Since Aluminum 26 doesn't last in quantity for much longer than a few million years, it must have been injected into the pre-Solar System nebula just before the nebula collapsed to form our system. In fact the supernova itself may have been responsible for that collapse.

The Early Solar System; Whether the initial collapse of the cloud is a result of a supernova outburst or some other mechanism, the next stage in the origin of the Solar System is fairly well understood. All free floating gas clouds in space are turbulent to one degree or another. This is especially true if an energy wave has been injected into the cloud by a supernova outburst. This turbulence will generally translate itself into a net rotation of the cloud. This rotational energy is expressed by a quantity called **angular momentum** and as gravity continues to collapse the cloud, the law of **conservation of angular momentum** predicts that the rotation rate of the cloud must increase. This is the same phenomenon which causes a skater's rotation rate to increase as he pulls in his arms or a tetherball's revolution speed to increase as it draws closer to its support pole.

As the cloud collapses due to gravity and spins faster to conserve angular momentum, its outer regions flatten along the equatorial plane. The material near the center of the cloud, less affected by the "flattening" centrifugal forces, contracts fastest and becomes the Sun (Figure 7.2).

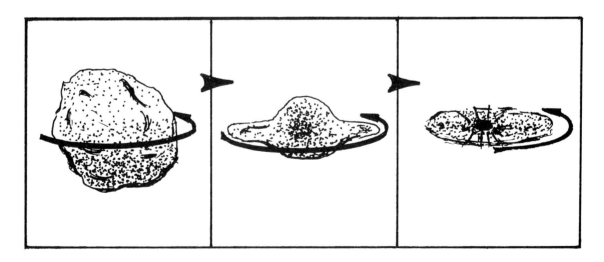

Figure 7.2: The Early Evolution of The Solar System

As the nebula collapsed, the energy released by the gravitational infall caused the cloud to heat up, particularly near the cloud's center where the temperature eventually reached 5 to 10 million Kelvins; the temperature required for the hydrogen fusion which powers the stars. The Sun was born.

In the disk surrounding the newly formed Sun, the temperature was a function of proximity to the cloud's center; 10 million K at the center, 2,000 K near the edge. The composition of the cloud was a mixture of primarily hydrogen and helium with quantities of iron, nickel, silicon, carbon and most other heavy elements thrown in.

As the cloud cooled, the gaseous material of which it was made condensed into solid particles. The condensation takes place in order of temperature. The atoms with the highest vaporization temperatures solidify first. These would be grains of aluminum and titanium oxide which condense at 1600 K. Since the temperatures and pressures were different in different parts of the nebula, we might expect the exact compositions of these regions to differ. This has been the case (regularities 5 and 6, page 152).

The Formation of Planets; In the outer nebula, the temperatures are low enough for water, ammonia and methane to condense. In the inner nebula, the temperatures kept these materials in the gaseous form so that they could be swept clear of the solar system by the strong solar wind of the forming sun. This explains why low density ices are most prominent in the outer solar system (regularity 8, page 152).

As the cloud cooled, more and more particles condensed out of the gaseous material, and as the particles suffered low speed collisions, they stuck together (either by electrostatic or gravitational forces) to form bodies called **planetesimals**. This process is called **accretion**. These bodies grew to several miles in diameter and some of them probably remain today as the asteroids in the asteroid belt.

Naturally, since these planetesimals form in the disk of the Solar nebula, they will all be orbiting around the equator of the Sun and in the same direction (regularities 1 and 2, page 151). Furthermore, since the disk of the nebula surrounds the equator of the Sun, the Sun's equator will lie parallel to the plane of the planetary orbits (regularity 3, page 151).

As time passed, planetesimals, colliding in low speed collisions, and held together by weak gravitational or electrostatic forces, grew in size to become the cores of the planets we see today. Planetesimals in elliptical orbits were now more likely to encounter other bodies and were preferentially absorbed. Planetesimals in circular orbits absorbed only those planetesimals which were at similar distances from the Sun. Computer models indicate that the coalescing planetesimals tend to form planets rotating in prograde directions. Planetesmals that formed and remained in the outer solar system are now called **comets**. Planetesimals that formed and remained in the inner Solar System (inside Jupiter's orbit) are now called **asteroids**.

According to Kepler's laws, the material closest to the Sun should orbit the Sun fastest and this will cause the nebula to divide into "rings" within which the planets will form.

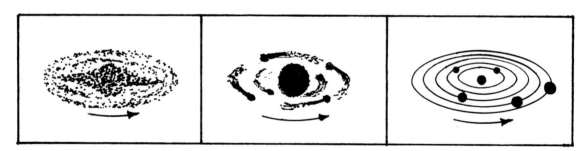

Figure 7.3: Formation of the Planets

In the inner Solar System, planetesimal accretion continued until Earth sized bodies were formed. It was too warm, however, for more volatile material (water for example) to solidify and become part of the newly forming planet. Farther out, volatiles could condense and become part of the planetesimals. This increased the mass of most of the "protoplanets" until their gravity was enough to start attracting material from the surrounding nebula. In this way, the outer planets became more massive than the inner planets and incorporated a large percentage of volatile material (hydrogen, helium, water, ammonia, methane).

The Formation of Satellites; As the massive planets formed, their condensation followed a sequence similar to the formation of the Solar System in general. Collapse of the massive protoplanet resulted in an increase in rotation and flattening of the system. Within the plane of this "miniature" Solar System, the major satellites of the outer planets formed from **accretion** (some satellites, such as Phoebe and Triton, probably formed by random capture) (Figure 7.4).

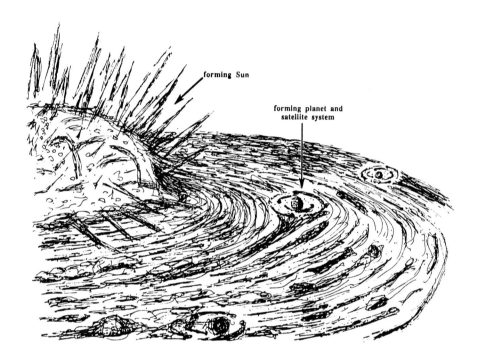

Figure 7.4: Formation of a Satellite System

The Formation of Asteroids; In the space between Mars and Jupiter, we find the asteroids. These bodies are probably planetesimals which could not condense because of the gravitational effects of Jupiter which increased the velocities of the planetesimals to the point where they could not stick together during a collision.

Further, it is believed that comets may be bodies, not incorporated into planetesimals, hurled by the gravity of the jovian protoplanets, into the Oort Cloud. Bodies hurled in the other direction fell into the Sun. Some of these may have been swept up by the Earth and may have provided some or all of the water in our oceans (regularity 9, page 152). Beyond this scenario, singular events were sure to occur. These events are almost certainly responsible for the retrograde rotation of Venus and the 98° obliquity of Uranus' rotation axis.

The picture painted here describes events which would be likely to take place during the formation of any single star system (perhaps as many as 25% to 35% of the star systems are single). If this is true, can we see such origins taking place? The answer seems to be yes! Studies made by infrared satellites have revealed star systems with disks of matter surrounding them. Could these be planetary systems in formation? Many believe that this is the case (Figure 7.5).

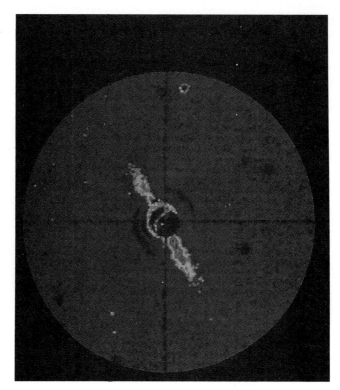

Figure 7.5: β Pictoris; A New Solar System?

APPENDIX I PLANETARY ORBITAL DATA

Planet	Distance from Sun (au)	(millions of miles)	Orbital Period	Orbit Eccentricity	Orbit Inclination to Ecliptic (degrees)
Mercury	0.39	36	88 d	0.206	7.0
Venus	0.72	67	225 d	0.007	3.4
Earth	1.00	93	365 d	0.017	0.0
Mars	1.52	142	687 d	0.093	1.8
Jupiter	5.20	484	11.9 y	0.048	1.3
Saturn	9.54	888	29.5 y	0.056	2.5
Uranus	19.18	1,786	84.1 y	0.046	0.8
Neptune	30.08	2,799	164.8 y	0.010	1.8
Pluto	39.52	3,666	248.6 y	0.248	17.1

APPENDIX II PLANETARY PHYSICAL DATA

Planet	Equatorial Diameter (miles)	Equatorial Diameter (Earth = 1)	Mass (Earth = 1)	Density (grams/cc)	Period of Rotation	Surface Gravity (Earth = 1)	Escape Velocity (miles/sec)	Atmosphere	Satellites	Rings
Mercury	3,031	0.38	0.055	5.5	58 d 14 h	0.38	2.7	essentially none	0	0
Venus	7,521	0.95	0.82	5.3	243 d (r)	0.90	6.4	carbon dioxide	0	0
Earth	7,926	1.00	1.00	5.5	23 h 56 m	1.00	6.9	nitrogen, oxygen	1	0
Mars	4,219	0.53	0.11	3.9	24 h 37 m	0.38	3.2	carbon dioxide	2	0
Jupiter	88,729	11.2	318	1.3	9 h 56 m	2.60	37.2	hydrogen, helium	16	1
Saturn	74,975	9.5	95.2	0.7	10 h 40 m	1.10	21.7	hydrogen, helium	18	thousands
Uranus	31,763	3.7	14.2	1.6	17 h 14 m	1.10	13.6	hydrogen, helium	15	11
Neptune	30,775	3.5	17.2	2.3	16 h 06 m	1.40	15.5	hydrogen, helium	18	5
Pluto	1,430	0.18	<.002	2.0	6 d 9 h (r)	0.06	0.6	methane	1	?

r = retrograde

APPENDIX III SATELLITES OF PLANETS

Planet	Satellite	Distance from Planet (x 10³ miles)	Diameter (miles)	Rotation Period (days)	Revolution Period (days)	Orbit Inclination (to Planetary Equator) (degrees)
Earth	Moon	238	2155	27	27	18 - 29
Mars	Phobos	5.8	17 x 12	0.32	0.32	1
	Deimos	14.3	9 x 7	1.26	1.26	3
Jupiter	J16 Metis	79	25	?	0.29	0
	J15 Adrastea	80	15 x 9	?	0.30	0
	J5 Amalthea	112	167 x 93	?	0.50	0
	J14 Thebe	138	74 x 56	?	0.67	0
	J1 Io	262	2,250	1.77	1.77	0
	J2 Europa	416	1,946	3.55	3.55	0
	J3 Ganymede	663	3,262	7.16	7.16	0
	J4 Callisto	1,167	2,976	16.69	16.69	0
	J13 Leda	6,878	5 (?)	?	239	27
	J6 Himalia	7,118	112	0.4	251	28
	J10 Lysithea	7,266	25	?	259	29
	J7 Elara	7,277	50	?	260	28
	J12 Ananke	13,144	19	?	631	147R
	J11 Carme	14,012	27	?	692	163R
	J8 Pasiphae	14,570	43	?	735	148R
	J9 Sinope	14,694	25	?	758	153R
Saturn	S18 Pan	83	12	?	0.58	0
	S15 Atlas	86	24 x 17	?	0.60	0
	S16 Prometheus	86	87 x 46	?	0.61	0
	S17 Pandora	88	68 x 41	?	0.63	0
	S11 Epimetheus	94	87 x 62	0.69	0.69	0
	S10 Janus	94	136 x 99	0.69	0.69	0
	S1 Mimas	115	244	0.94	0.94	1
	S2 Enceladus	145	311	1.37	1.37	0
	S3 Tethys	183	650	1.89	1.89	1
	S13 Telesto	183 *	16 x 7		1.89	0
	S14 Calypso	183 *	19 x 10		1.89	0
	S4 Dione	233	693	2.74	2.74	0
	S12 Helene	233 †	22 x 12	?	2.74	0
	S5 Rhea	327	9.47	4.52	4.52	0
	S6 Titan	758	3,193	?	15.94	0
	S7 Hyperion	918	217 x 124	chaotic	21.28	0
	S8 Iapetus	2,207	890	79.33	79.33	15
	S9 Phoebe	8,017	143 x 130	0.4	550.5	150R

* S13, S14 in lagrangian points of Thethys' orbit
† S12 in lagrangian point of Dione's orbit
R = retrograde orbit

SATELLITES OF PLANETS (continued)

Planet	Satellite	Distance from Planet (x 10³ miles)	Diameter (miles)	Rotation Period (days)	Revolution Period (days)	Orbit Inclination (to Planetary Equator) (degrees)
Uranus	U6 Cordelia	31	16	?	0.34	0
	U7 Ophelia	33	20	?	0.38	0
	U8 Bianca	37	27	?	0.44	0
	U9 Cressida	38	41	?	0.46	0
	U10 Desdemona	39	36	?	0.48	0
	U11 Juliet	40	52	?	0.49	0
	U12 Portia	41	68	?	0.52	0
	U13 Rosalind	43	36	?	0.56	0
	U14 Belinda	47	42	?	0.62	0
	U15 Puck	53	99 x 93	?	0.76	0
	U5 Miranda	81	300	1.41	1.41	3
	U1 Ariel	119	719	2.52	2.52	0
	U2 Umbriel	166	738	4.14	4.14	0
	U3 Titania	272	992	8.71	8.71	0
	U4 Oberon	363	961	13.46	13.46	0
Neptune	1989N6	30	32	?	0.30	4
	1989N5	31	50	?	0.31	~0
	1989N3	33	93	?	0.33	~0
	1989N4	38	111	?	0.40	~0
	1989N2	46	918	?	0.55	~0
	1989N1	73	248	?	1.12	~0
	N1 Triton	219	1,677	5.88	5.88	159R
	N2 Nereid	3,419	248	?	260.2	28
Pluto	P1 Charon	12	738	6.39	6.39	

* S13, S14 in lagrangian points of Thethys' orbit
† S12 in lagrangian point of Dione's orbit
R = retrograde orbit

APPENDIX IV LEARNING ACTIVITY 1

THE GEOCENTRIC HYPOTHESIS AND PLANETARY MOTION

The geocentric model of Ptolemy allows for the prediction of planetary motion, including retrograde motion (p. 9, Figure 2.3). In this exercise you will use a geocentric epicyclic diagram to predict the time that it will take Venus to go from **superior conjunction** (on the other side of the Sun as seen from the Earth) to **inferior conjunction** (directly between the Earth and the Sun).

On the next page you will find an **epicyclic diagram** for the planet Venus. At the center of the diagram is the Earth (point **E**). At a radius of 1.00 au is the **deferent** on which point **S** orbits the Earth with a period of 1.00 years. Since there are 360° in a circle, S moves about E at a rate of about 1°/day.

Centered on point **S**, with a radius of 0.72 au, is the **epicycle** on which the planet Venus (point **V**) rides. V orbits S with a period of 0.62 years, for a rate of about 1.6°/day. See figure below.

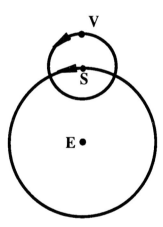

A. Determine the time that it takes Venus to go from superior conjunction to inferior conjunction. Start with Venus at superior conjunction and move S and V in 25-day increments until **V** is between **S** and **E**.

B. The time from superior conjunction to the next superior conjunction is just twice the time it takes to go from superior conjunction to inferior conjunction. The time period from superior conjunction to superior conjunction is known as the **synodic period**. What is the synodic period of Venus?

163

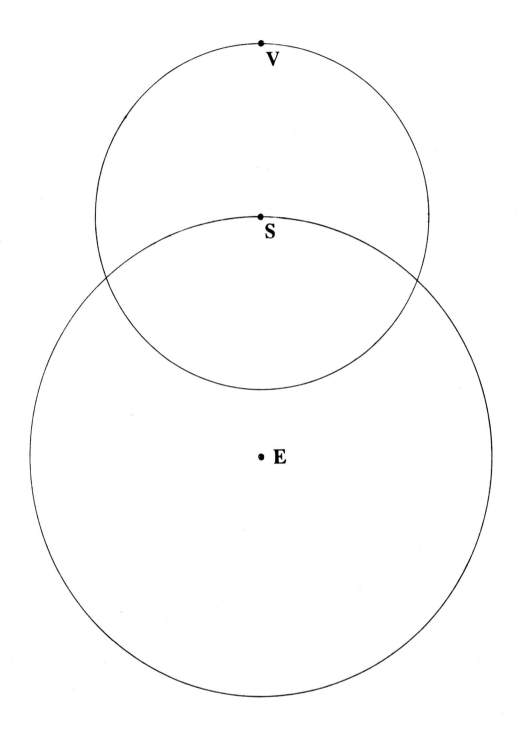

APPENDIX IV LEARNING ACTIVITY 2

RETROGRADE MOTION IN HELIOCENTRIC ORBITS

The purpose of this exercise is to demonstrate the phenomenon which gave the proponents of the geocentric hypothesis so much difficulty; **retrograde motion**. This apparent change in the direction of motion of the planets was described by the epicycles of Ptolemy (pp. 9-10 and Learning Activity 1). In the 16th century, however, Copernicus revived the heliocentric hypothesis and explained retrograde motion as a result of the different speeds of the planets in their orbits. This exercise demonstrates this effect (see diagram on p. 11).

In the chart below are tabulated positional data for Earth and Mars. The positions are given in terms of heliocentric celestial longitude (angular distances from the vernal equinox, along the ecliptic).

The problem: Determine the dates when Mars: a) starts moving retrograde (westward) and, b) resumes prograde (eastward) motion. These dates are called **stationary points**.

To accomplish the solution to this problem, graph the data from the chart on the polar graph found on the next page. Label each planetary position with the appropriate date. Draw in the lines of sight from the Earth to Mars for each date. As long as the line is "twisting" counterclockwise, the motion of Mars as seen from Earth is **prograde**. When two successive lines become parallel, Mars is **stationary**. When the line starts twisting clockwise, Mars is moving retrograde. The next set of parallel lines indicates the second **stationary interval**. This is followed by counterclockwise twisting (prograde motion).

Date	Earth	Mars	Date	Earth	Mars
Feb 28	320°	20°	Jun 30	80°	80°
Mar 20	340°	30°	Jul 20	100°	90°
Apr 10	0°	40°	Aug 10	120°	100°
Apr 30	20°	50°	Aug 30	140°	110°
May 20	40°	60°	Sep 20	160°	120°
Jun 10	60°	70°	Oct 10	180°	130°

a. First stationary interval; from _____ to _____

b. Secondary stationary interval; from _____ to _____

c. Interval of retrograde motion; from _____ to _____

d. Mark the stationary retrograde intervals on the graph on page 166.

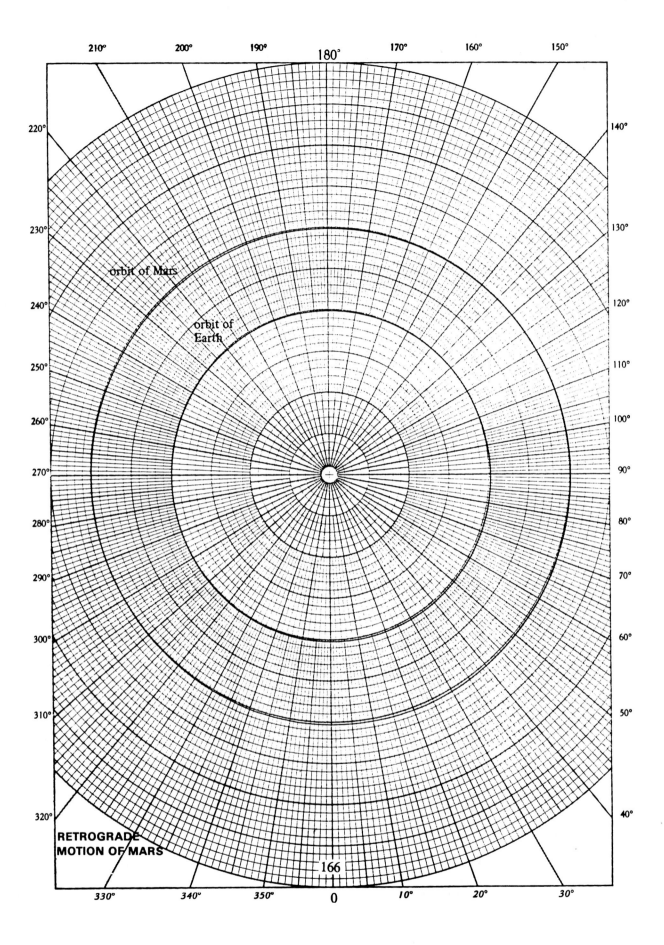

RETROGRADE MOTION OF MARS

APPENDIX IV LEARNING ACTIVITY 3

KEPLER'S FIRST LAW; THE ELLIPSE

The purpose of this exercise is to demonstrate the properties of the ellipse; the geometrical shape which defines the planetary orbits (page 12). On the graph paper found on the next page, construct an ellipse of eccentricity = 0.3 and major axis = 15 cm. The eccentricity of an ellipse is defined as;

$$e = \frac{\overline{FF'}}{2a}$$

where $\overline{FF'}$ = the distance between the foci
and $2a$ = the length of the major axis

Place 2 tacks on the vertical line a distance $\overline{FF'}$ apart (Figure 1). Connect these two tacks with a string of length 2a. Use the string as a guide to draw the ellipse as shown below.

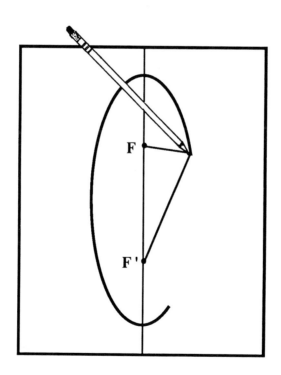

Construction of an Ellipse

167

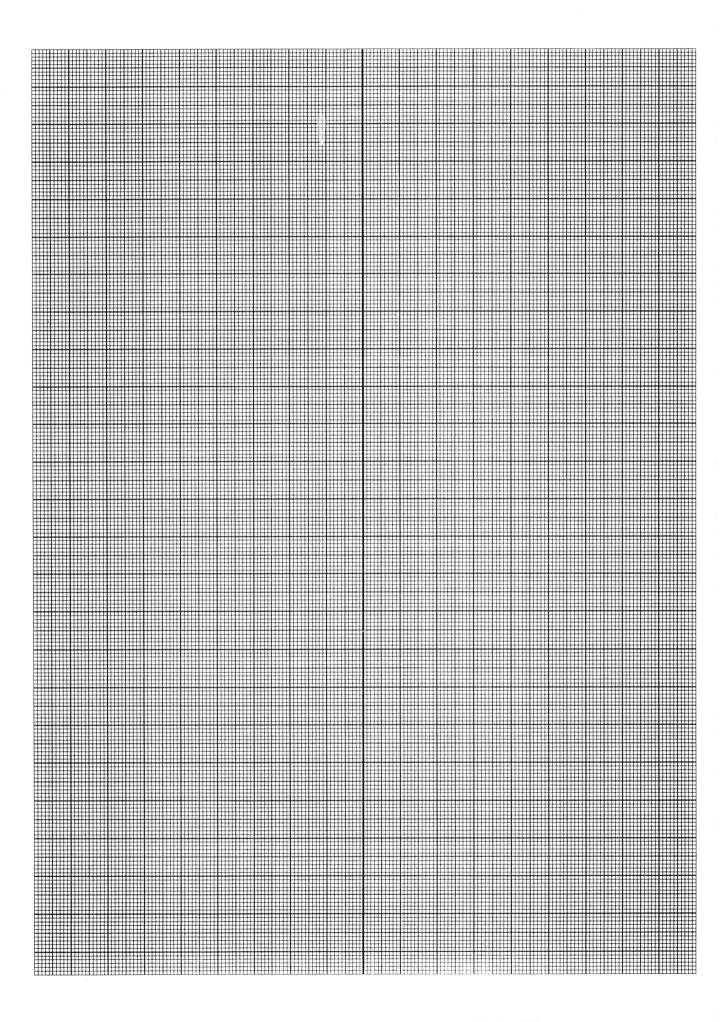

APPENDIX IV   LEARNING ACTIVITY 4

KEPLER'S SECOND LAW; LAW OF AREAS

Using the ellipse constructed in Learning Activity 3, choose one focus as the location of the Sun. Place the planet at its aphelion point (farthest point from the Sun). Let the heliocentric longitude of the aphelion point be 0°. Assume that after 10 weeks the planet has a heliocentric longitude of 15°. Using Kepler's 2nd law, determine how many weeks must elapse (starting from aphelion) until perihelion is reached (see figure below). What is the period of revolution of the planet? (Hint: Count the number of boxes enclosed by the 15° angle and the arc of the ellipse. Then count the number of boxes between the major axis and the arc of the ellipse. Since you know that it took 10 days to sweep out 15° worth of area, you can compute how long it should take to sweep out 180° worth of area.)

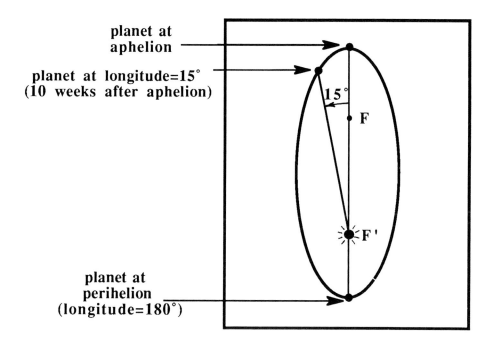

Determination of Period from Kepler's 2nd Law

APPENDIX IV   LEARNING ACTIVITY 5

KEPLER'S THIRD LAW, HARMONIC LAW

A. Using the planetary period determined in Learning Activity 4, determine the semi-major axis of the planetary orbit (this is the average Sun-planet distance). To do this you will use the third law.

$$a^3 = p^2$$

where a = average Sun-planet distance in centimeters*

and p = planetary revolution period in years

To solve for **a**, you will need to take a cube root. This can be done on a calculator or by successive approximations. An exact answer is not required; a "ballpark" figure will do.

B. Using Kepler's 3rd Law ($a^3 = p^2$), determine the periods of revolution of the major planets in the Solar System. The semi-major axes (a) of the planetary orbits are given in the table below. Compare your answers to the accepted answers (Appendix I).

Planet	a (au)	p (years)
Mercury	0.39	
Venus	0.72	
Earth	1.00	
Mars	1.52	
Jupiter	5.20	
Saturn	9.54	
Uranus	19.18	
Neptune	30.08	
Pluto	39.52	

*In this exercise, 1 cm = 1 au.

APPENDIX IV　　　　　　　LEARNING ACTIVITY 6

SURFACE GRAVITY

Newton's Law of Gravity (p. 14) allows one to compute the surface gravity of a planet knowing only the planet's mass and radius. In particular, if the mass and radius are known in **earth masses** and **earth radii**, the equation for surface gravity (in **earth gravities**) simplifies to;*

$$g = \frac{M}{R^2}$$

where　g = surface gravity of planet in "earth gravities"

M = mass of planet in "earth masses"

and　R = radius of planet in "earth radii"

Using the above equation, compute the surface gravities on the following planets.

Planet	Mass (earth masses)	Radius (earth radii)	Surface Gravity (earth gravities)
Mercury	0.05	0.38	
Mars	0.1	0.5	
Jupiter	318	11.2	
Saturn	95	9.5	
Uranus	14	3.7	
Neptune	17	3.5	

*This equation can be derived by dividing Newton's Law of Gravity for your weight on the planet by Newton's Law of Gravity for your weight on Earth.

APPENDIX IV   LEARNING ACTIVITY 7

THE ORBIT OF MERCURY

The orbit of the planet Mercury has the second greatest eccentricity of any major planetary orbit in our Solar System. In this activity you will determine the value of this eccentricity. Table 1 gives greatest western and eastern elongations for the planet Mercury. Greatest elongation refers to the maximum angular separation between a planet (Mercury or Venus) and the Sun (Figure 1).

Date	Greatest Western Elongation	Date	Greatest Eastern Elongation
February 16, 1976	26°	April 28, 1976	21°
June 15, 1976	23°	August 26, 1976	27°
October 7, 1976	18°	December 20, 1976	20°
January 29, 1977	25°	April 10, 1977	19°
May 28, 1977	25°	August 8, 1977	27°
September 21, 1977	18°	December 3, 1977	21°
January 11, 1978	23°	March 24, 1978	19°
May 9, 1978	26°	July 22, 1978	27°
September 5, 1978	18°	November 16, 1978	23°
December 24, 1978	22°	March 8, 1979	18°
April 21, 1979	27°	July 3, 1979	26°
August 19, 1979	19°	October 29, 1979	24°
December 7, 1979	21°	February 19, 1980	18°
April 2, 1980	28°	June 14, 1980	24°
August 1, 1980	19°	October 11, 1980	25°
November 19, 1980	20°	February 2, 1981	18°
March 16, 1981	28°	May 27, 1981	23°
July 14, 1981	21°	September 23, 1981	26°
November 3, 1981	19°	January 16, 1982	19°
February 26, 1982	27°	May 9, 1982	21°
June 26, 1982	22°	September 6, 1982	27°

Table 1

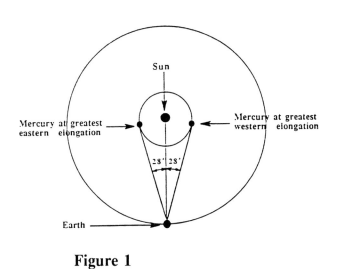

Figure 1

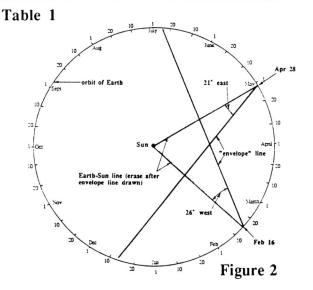

Figure 2

A. Using the data from Table 1 and the chart on page 177, graph the greatest western and eastern elongations of Mercury as shown in Figure 2. This will give you the **envelope** of the orbit of Mercury.

B. Use a french curve (if available) to draw in the elliptical orbit (Figure 3).

C. Find the **perihelion point** of the orbit (point closest to the Sun). Draw in the major axis by connecting the perihelion point with the Sun and continuing the line until it hits the **aphelion point** (point farthest from the Sun) (Figure 3).

D. Find the 2nd focus of the ellipse (the first focus is where the Sun is located). It is located on the major axis, as far from the aphelion point as the first focus is from the perihelion point.

E. Compute the **eccentricity** of the orbit of Mercury. The eccentricity of an ellipse is given by:

$$e = \frac{\overline{FF'}}{2a}$$

where $\overline{FF'}$ = the distance between the foci

and $2a$ = length of the major axis (a = semi major axis)

F. Compare your answer with the correct value found in Appendix I. Compute your % error. [% error = 100 x (your answer - correct answer) ÷ correct answer]

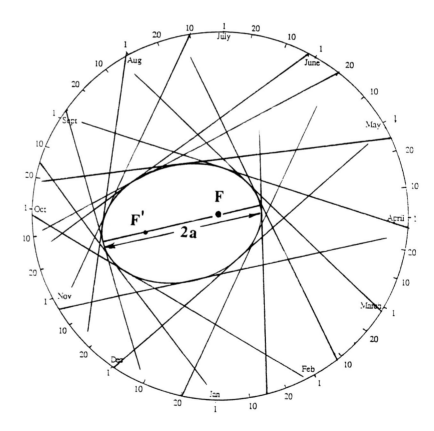

Note: This diagram is drawn neither at the correct scale nor with the correct data. It is just an example.

Figure 3

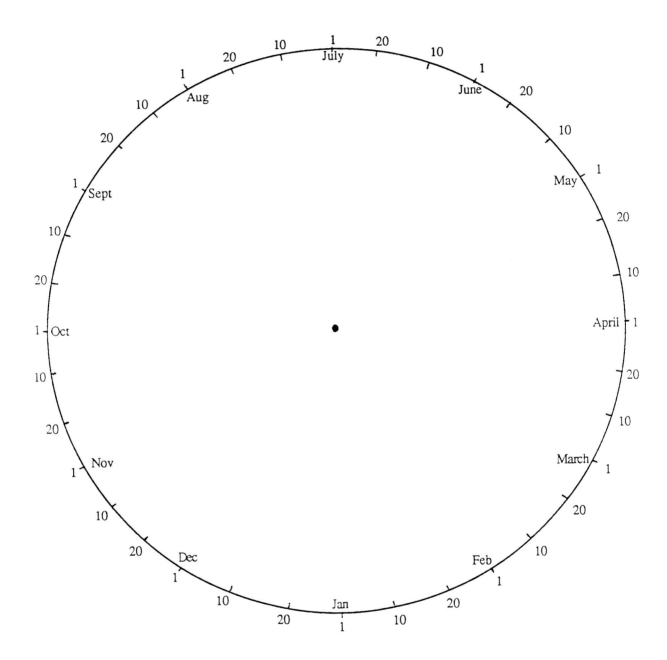

APPENDIX IV   LEARNING ACTIVITY 8

THE SEMI MAJOR AXIS OF VENUS AND THE DETERMINATION OF THE AU

In the early 16th century, Nicolaus Copernicus (1473-1543) was able to create a "scale model" of the Solar System by observing the planets as they moved in the sky. In this model, he was able to place the planets at their correct <u>relative</u> distances from the Sun. The unit of distance used in the model was the average Earth-Sun distance (which we, today, call the **astronomical unit**), although he did not know how big this unit was. We will use the planet Venus to demonstrate how Copernicus created his model.*

The angle between the Sun and Venus, as seen from Earth, is called the **elongation angle** (See figure below). For Venus, the maximum possible elongation angle is 48°. Notice that at

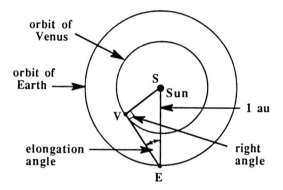

maximum elongation, the angle EVS <u>must be</u> 90°! At maximum elongation, then, we have a triangle for which three things are known; 1 side (**ES**=1) and 2 angles (**EVS** = 90° and **SEV** = 48°)

A. The Semi-Major Axis of the Orbit of Venus in Au's; Using trigonometry [sine (SEV) = VS/ES] or construction (a ruler and protractor), determine the semi major axis of the orbit of Venus (side **VS**). The units of your answer will be au's. The determination of the distance of Venus from the Sun in <u>miles</u> requires that we know the length of the au in miles. One method of doing this is to measure the distance to Venus in miles (by radar, for example) at the time that you know its distance in au's. Then, a simple proportion will allow you to determine the number of miles in an au.

B. Determination of the Au; Using the triangle from part A, determine [by trigonometry (cosine), or construction] the value of **EV** in au's. Assume that on the night in question (maximum elongation) the Earth/Venus distance is determined (by radar) to be 62,229,000 miles. Set up a proportion and determine the number of miles in one au.

C. Determine the semi-major axis of the orbit of Venus in <u>miles</u>. Compare this value to that given in Appendix I. Compute your % error.

*His method for Mercury and Venus is easy to understand, while the superior planets (planets whose orbits lie beyond Earth's) require a more complex method.

APPENDIX IV   LEARNING ACTIVITY 9

THE MASS OF MARS

We saw on page 12 that Kepler's 3rd law of planetary motion is given by the expression $p^2=ka^3$ where **p** is the sidereal period of the planet, **a** is the planet's mean distance from the Sun and **k** is a constant whose value depends on the unit system chosen (e.g. if p is in years and a is in au's*, then k=1). In the mid 17th century, Isaac Newton developed the law of gravity and demonstrated that Kepler's laws of planetary motion could be derived from that law. He further discovered that Kepler's 3rd law was slightly more complex than Kepler believed. Newton's version of Kepler's 3rd law (which is a more complete and general version than Kepler's) is given by;

$$(m_1+m_2)\, p^2 = ka^3$$

Where m_1 and m_2 are the masses of the two orbiting bodies and **p**, **a** and **k** are as previously defined. The unit of mass depends on what units are chosen for **p** and **a**. For example, if **p** and **a** are in years and au's, **k** will equal 1 and mass will be in units of the mass of the Sun (**solar masses**). Another unit system for which k=1 are units of **earth masses**, **earth radii** (1 earth radius=3950 miles) and **canonical time units** (the **ctu** is the period of revolution of an Earth satellite at a distance of 1 **earth radius** from the Earth's center. 1 **ctu** = 84.5 minutes). Using these units and Newton's version of the 3rd law, it is possible to determine the mass of a planet by observing the orbit of one of its small satellites.† The form of the equation used will be:

$$m_{planet} + m_{satellite} = \frac{a^3}{p^2} \quad (k=1 \text{ if proper units are used})$$

Problem: Determine the mass of Mars in earth masses given the following observations (do the problem with both satellites and compare your answers);

Martian satellite	P	A	Mass of Mars
Phobos	460 minutes	5,828 mi	
Deimos	1.26 days	14,260 mi	

*The "au" (astronomical unit) is a unit of distance equal to the mean Earth-Sun distance. 1 au = 93,000,000 miles.
†If a small satellite is used, $m_{satellite}$ may be ignored due to it being negligible.

APPENDIX IV   LEARNING ACTIVITY 10

THE MASS OF THE PLUTO-CHARON SYSTEM

As seen in Learning Activity 9, the mass of a planet/satellite system can be calculated by Newton's version of Kepler's 3rd law;

$$m_{planet} + m_{satellite} = k \frac{a^3}{p^2}$$

where:
- m_{planet} = mass of planet
- $m_{satellite}$ = mass of satellite
- a = average planet-satellite distance
- p = period of revolution of satellite
- k = constant of proportionality

The value of **k** depends on the units chosen. For certain "unit sets" **k** = 1. We have already seen that this is true for **solar masses, astronomical units** and **years** and also for **earth masses, earth radii** and **ctu's**. A third unit set for which **k** = 1 is: **earth masses, earth-moon distances** and **sidereal months** (1 earth-moon distance = 240,000 miles and 1 sidereal month = 27d 7h = 27.3 days).

The problem; Given that the Pluto/Charon distance is 12,200 miles and the period of revolution of the Pluto/Charon system is 6.4 days, determine the mass of the Pluto/Charon system. Remember that one earth radius = 3,950 miles and one ctu = 84.5 minutes. Also remember to state proper units for your answer.

$$m_{Pluto} + m_{Charon} = \frac{a^3}{p^2} = \underline{\qquad} ?$$

APPENDIX IV   LEARNING ACTIVITY 11

RADIANT POINT OF METEOR SHOWER

About ten times per year, the Earth passes through the orbit of a defunct comet. This produces the phenomenon known as a **meteor shower** (p. 90). Because the meteoroid particles are moving along parallel paths through space, the meteor tracks, when backtracked, appear to radiate from a point in the sky; the **radiant point**. The radiant point gives the shower its name. The figure on the next page shows a region of the winter sky and the meteor tracks seen during an hour of observing. Backtrack these tracks and determine: a) the radiant point (name of the constellation in which the radiant is found), b) the star nearest the radiant point and, c) the date of the shower (from Figure 4.18).

_____ Radiant constellation

_____ Radiant star

_____ Date of shower

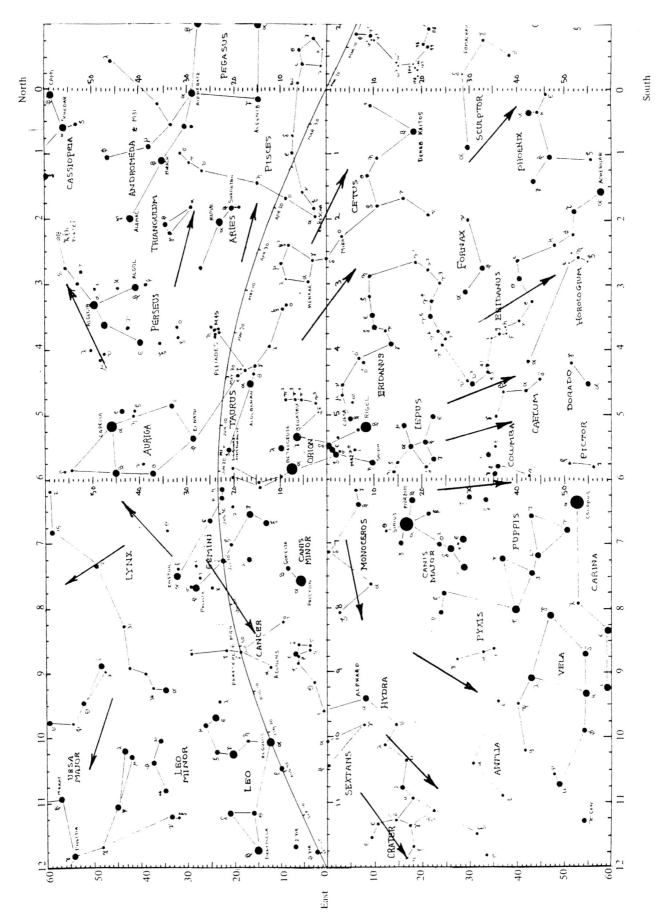

APPENDIX IV   LEARNING ACTIVITY 12

TIME FROM LUNAR PHASES

The phases of the Moon are described on pages 97 - 100 and again in the figure below. An interesting aspect of the lunar phases is that it is possible to tell the time of day from them. It is obvious, for example, that since the full moon is opposite the Sun in the sky, it must rise at sunset and set at sunrise. Seeing the full moon overhead, therefore, indicates that it is midnight. Since the first quarter moon is 90° (=6h) east of (behind) the Sun, it will rise at noon and set at midnight. Seeing the first quarter moon overhead, therefore, means that it is sunset.

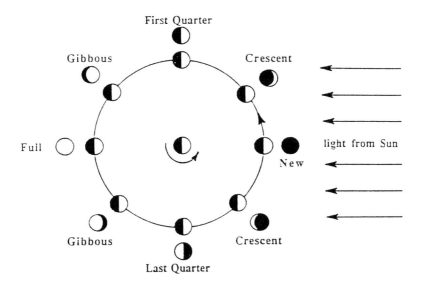

Use Figure 1 to answer the following questions about time:

A. What is the time of day at which the new moon rises?

B. What is the time of day if the last quarter moon is setting?

C. You notice the new moon is overhead (can you actually do this). What time is it?

D. What time is it if a three-day old moon is setting?

E. What time will a 15-day old moon rise?

F. What is the time of day at which the last quarter moon is overhead?

G. What time will a 7-day old moon rise?

APPENDIX IV   LEARNING ACTIVITY 13

THE SUNSPOT CYCLE

The sunspot cycle of our Sun was discussed in Chapter 6 (pp. 141 - 143). The purpose of this Learning Activity is to use past sunspot data to predict future sunspot maxima and minima. Figure 1 and Figure 2, below, show a butterfly diagram and sunspot number graph for past years.

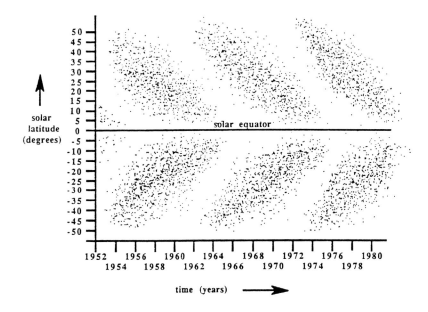

Figure 1: Butterfly Diagram

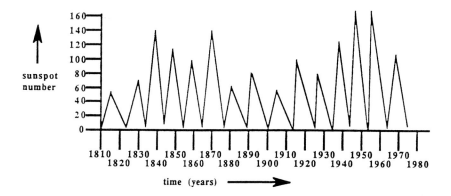

Figure 2: Sunspot Number Graph

Using the graphs on the previous page and a mean sunspot cycle period of 11 years, predict the next three maxima and minima which will occur <u>after</u> 1980.

                                                Date

Maximum   _____

Minimum   _____

Maximum   _____

Minimum   _____

Maximum   _____

Minimum   _____

APPENDIX V

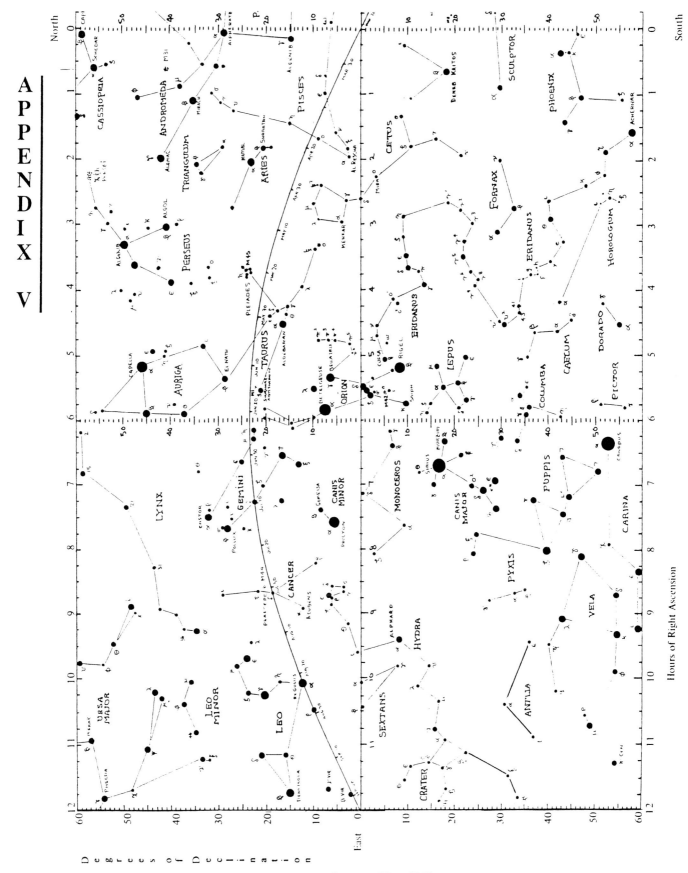

Equatorial Star Chart (Fall/Winter)

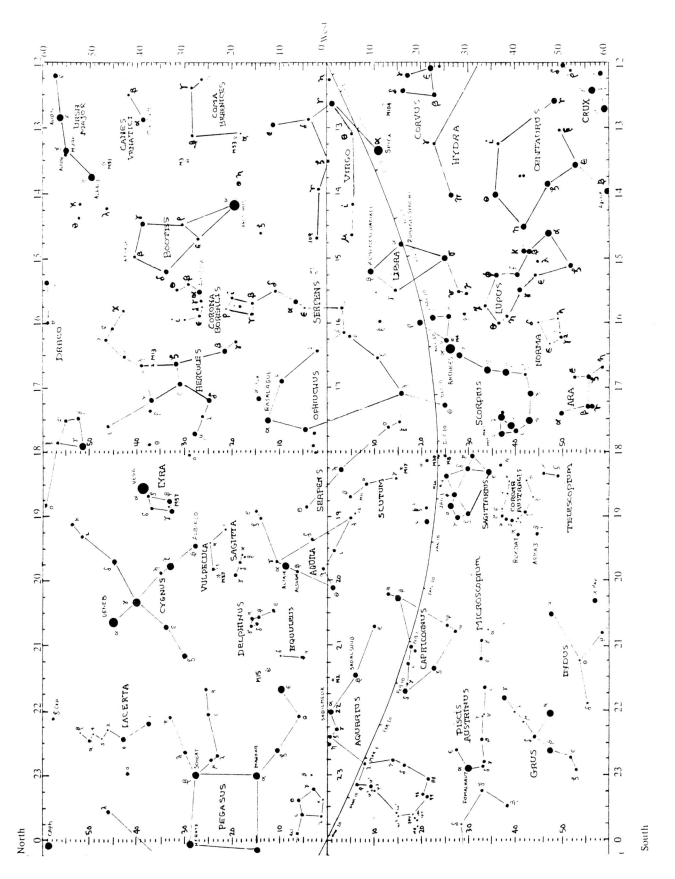

Equatorial Star Chart (Spring/Summer)

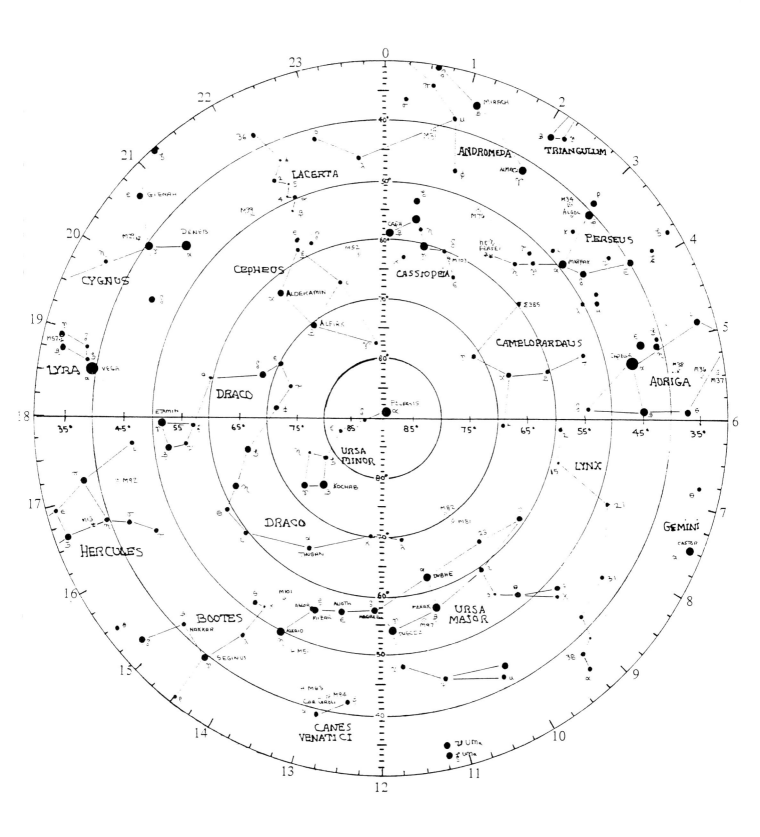

North Polar Star Chart

PHOTO CREDITS

California Institute of Technology, Hale Observatories
Figures 1.2, 1.6, 3.21, 3.22, 3.23, 3.41, 3.56, 3.83, 4.7, 5.6, 5.8, 5.9, 5.17, 5.19, 5.20, 5.22, 5.44, 6.1, 6.6, 6.9, 6.10, 6.11, 6.18, 6.19, 6.20, 6.23, 6.24, 6.25, 7.1

NASA/JPL
Figures 1.5, 3.6, 3.7, 3.9, 3.13, 3.14, 3.15, 3.16, 3.17, 3.18, 3.24, 3.26, 3.27, 3.28, 3.29, 3.30, 3.31, 3.32, 3.33, 3.34, 3.35, 3.36, 3.37, 3.38, 3.39, 3.40, 3.42, 3.43, 3.44, 3.46, 3.47, 3.48, 3.49, 3.50, 3.51, 3.52, 3.53, 3.54, 3.55, 3.57, 3.58, 3.59, 3.60, 3.61, 3.63, 3.64, 3.65, 3.66, 3.67, 3.68, 3.69, 3.70, 3.71, 3.73, 3.76, 3.77, 3.78, 3.79, 3.80, 3.81, 3.85, 5.21, 5.23, 5.24, 5.25, 5.26, 5.33, 6.22, 7.5

Figure 2.6 - Mysterium Cosmographicum, "Johannes Kepler"

Figure 3.11 - Soviet Space Agency

Figure 3.12, 3.25 - NASA/JPL/USGS

Figure 3.19 - G. Sciaparelli

Figure 3.20 - Lowell Observatory Photograph

Figure 3.84 - U.S. Naval Observatory/J.W. Christy/U.S. Navy Photograph

Figure 4.6 - European Space Agency

Figure 4.10 - Photo by Author

Figure 4.19 - Meteor Crater; Northern Arizona

Figure 4.20 - Birmingham News Co.

Figure 5.7, 5.18 - Lick Observatory Photograph

Figure 5.16 - Galileo Galilae

Figure 5.30 - T.V. Image

Figure 5.41, 5.42 - John Mottman

Figure 5.43 - R. L. Smith

Figure 6.8 - Adapted from Anders and Ebihara (1982)

Cover Photo - Photo by Author

Appendices I, II, III , Figure 6.2 - Adapted from "Astronomy: The Cosmic Journey", Hartmann

INDEX

A

Accretion theory, 121
Adams, John Couch, 72
Aldrin, "Buzz", 117
Alpha Centauri, 5
Amalthea, 57
Anaxogoras, 9
Angular momentum, conservation of, 153
Apollo asteroids, 83
Apollo series, 116-118
Ariel, 70
Aristarchus, 2, 10
Aristotle, 9
Armstrong, Neil, 117
Asteroid belt, 81
Asteroids, 4
Astronomical unit, 4

B

Babcock theory of sunspots, 143
Barycenter, 103
Beta Pictoris, 155
Bode-Titius progression, 82
Bolides, 90
Brahe, 2, 11

C

Callisto, 55, 57
Capture theory, 120, 121
Cassini's Division, 63
Ceres, 4, 81-83
Charon, 77, 78
Chiron, 84
Christy, James, 77
Collins, Michael, 117
Columbia spacecraft, 117
Comets, 4, 84-90
    Comet West, 88
    formation of tails, 85-88
    head, 85
    naming of, 89
    nucleus, 85
    Oort Cloud, 89, 90, 155
    orbits, 85
    origin, 89
    short period, 85
    structure, 85
    tail, 85-89
Constellations, 7
Copernicus, 2, 10
Coronagraph, 144

D

Dark line spectrum, 136
Deferents, 9
Deimos, 45, 46
Democritus, 9
Differential rotation, 143
Dione, 65
Draconistic month, 128
Dust tail (of comet), 87, 88

E

Eagle spacecraft, 117
Earth, 3
    circumference, 10
Eclipses, 122-129
    annular solar, 124
    central eclipses, 129
    eclipse cycles, 128
    eclipse seasons, 128
    Einstein effect, 126
    lunar, 126, 127
    partial lunar, 127
    partial solar, 123, 124
    path of totality, 125, 126
    penumbral lunar, 127
    saros, 128
    shadows, 122
    solar, 123
    total lunar, 127
    total solar, 123, 125, 144
    umbral, 127
    visibility, 129
Ecliptic 96
Einstein effect, 126
Electromagnetic radiation, 134-137
Electromagnetic spectrum, 134-135
Enceladus, 64, 65
Epicycles, 9
Eratosthenes, 2, 10
Escape velocity, 112
Eudoxus, 9
Europa, 52, 54

F

Fireballs, 90
Fission theory, 120
Free-fall, 16

G

Galaxies, 5
Galilean satellites, 47
Galileo, 2, 12
Galileo, telescopic observations, 12, 13
Gamma rays, 134
Ganymede, 55, 56
Gas tail (of comet), 86, 87

Gemini series, 116
Giotto, 86
Gravity, 3
Greek astronomy, 2

H

Halley's comet, 85
Halley, Edmund, 84
Helium, 144
Heraklides, 9
Herschel, Wm., 67
Hipparchus, 10
Hodges, Ermeline, 94
Hubble space telescope, 78
Huygens, Christian, 59
Hyperion, 65, 66

I

Iapetus, 66
Impact theory, 121, 122
Infrared radiation, 134
Io, 52, 53

J

Jovian Planets, 47
Juno, 83
Jupiter, 13, 47
    Amalthea, 57
    atmosphere, 49, 51
    belts and zones, 47
    composition, 47, 48
    core, 51
    density, 48
    diameter, 47
    distance from Sun, 47
    Galilean satellites, 13, 47, 52
    internal heat, 51
    internal structure, 51
    mass, 48
    orbit, 47
    period of revolution, 48
    period of rotation, 48
    pressure, 51
    red spot, 49, 50
    ring, 58
    satellites, 52, 57
    spaceprobes, 48
    temperature, 51

K

Kepler, 2, 11
    1st law, 11
    2nd law, 12
    3rd law, 12
Kirkwood gaps, 84
Kuiper Airborne Observatory, 69

L

Lagrangian points, 84
Laws of nature, 1, 8
Leverrier, Urbain, 72
Light day, 4
Light hour, 4
Light minute, 4
Light second, 4
Light year, 4
Lines of nodes, 97
Lowell, Percival, 32, 76
Lunar orbiter, 115, 116

M

Magellan radar mapper, 28-34
Mariner 10, 26
Mars, 32
    appearance, 34
    axial tilt, 34
    canali, 32
    diameter, 34
    distance from Sun, 35
    Lowell's "canals", 32
    Mariner probes, 35
    Mt. Olympus, 36, 39, 40
    opposition, 34
    outflow channels, 42, 43
    polar caps, 34, 42, 44
    revolution, 34
    rotation, 34
    runoff channels, 42, 43
    satellites, 45, 46
    surface, 36
    Tharsis plateau, 36, 39
    Valles, Marineris, 41, 42
    Viking, 35, 44, 45
    volcanoes, 36, 39, 40
    water on, 42
Maunder minimum, 142
Maunder, E. W., 141
Mercury, 17
    albedo, 20
    atmosphere, 21
    craters, 21
    density, 20
    diameter, 20
    Earth-based observations, 20
    elongation, 17
    magnetic field, 21
    mass, 20
    orbit, 17
    perihelion advance, 22
    rotation, 17-20
    scarps, 20
    series, 116
    surface temperature, 21

  test for general relativity, 22
Meteor Crater, 93, 94
Meteor showers, 90
  origin, 91-93
  radiant point, 90
Meteorites, 93
  achondrites, 94
  carbonaceous chondrites, 94
  chondrites, 94
  iron, 94
  stony, 94
  stony-iron, 94
Meteoroids, 90
Meteors, 90
  origin, 91
  sporadic, 90
Microwave radiation, 134
Milky Way galaxy, 5
Mimas, 64
Minor planets, 81
Miranda, 70
Moon, 13, 95-131
  albedo, 98
  Apollo, 104
  atmosphere, 112
  core, 119
  craters, 105-111
  delay in moonrise, 100
  diameter, 102, 103
  distance from Earth, 95, 101
  escape velocity, 112
  far side, 120
  highlands, 105-110, 118
  interior, 119
  laser ranging, 101
  lowlands, 105-110, 119
  maria, 104, 105
  mass, 103, 104
  moonrocks, 104
  mountains, 105-110
  nodes, 97
  orbit, 95
  origin, 120, 121
  phases, 97-100
  physical properties, 101-103
  regolith, 117
  rilles, 111
  Roche's limit, 131
  rotation, 102
  sidereal and synodic Month, 96
  space program, 113-118
  surface gravity, 112
  surface temperature, 112
  surface water, 112
  surface, 104-112
  terminator, 112
  triangulation, 101

N
Neptune, 72
  clouds, 72
  density, 72
  diameter, 72
  discovery, 72
  distance from Sun, 72
  Galileo's observations. of, 72
  gravity, 72
  Great Dark Spot, 73
  magnetic field, 73
  mass, 72
  orbit, 72
  period of revolution, 72
  period of rotation, 73
  rings, 74
  Roche's limit, 74
  satellites, 74
  structure, 72
  temperature, 72
  Triton, 155
  Voyager, 73
Nereid, 75
Newton, Isaac, 2, 13
  contributions, 14
  law of gravity, 14
  laws of motion, 14
Nodes, 97

O
Oberon, 70
Oort cloud, 89, 90, 155
Orbits, 15
Origin of Solar System, 151-155
  accretion, 154, 155
  satellites, 155

P
Pallas, 83
Parmenides, 9
Penumbra, 122, 138, 140
Phases of Moon, 97-100
Phobos, 45, 46
Phoebe, 66, 155
Piazzi, Giuseppí, 81
Planck's law, 135
Planetesimals, 84
Planetoids, 4
Planets, 7
  diameter, 3
  major planets, 3
  minor planets, 4
  orbits, 3
Plato, 2, 9, 76
Pluto
  atmosphere, 80
  discovery, 76

discovery of Charon, 77
eclipse seasons, 78
HST view, 78
mass, 78
orbit, 76
period of rotation, 77
period of revolution, 76
possible origin, 80
structure, 80
surface, 78 - 80
Ptolemy, 2, 10
Puck, 70
Pythagoras, 2, 9

R

Radar astronomy, 18
Radio waves, 134
Ranger, 113-114
Rays, 20
Region of transit, 122
Regression of lines of nodes, 97
Relativity, theory of, 22, 126
Renaissance astronomy 2, 10
Retrograde motion, 7, 10
Rhea, 65
Rilles, 111
Roche's limit, 63, 74, 131

S

Satellites, 4, 155
Saturn V rocket, 116
Saturn, 13, 59
    atmosphere, 62
    Cassini's division, 59
    density, 59
    diameter, 59
    interior, 62
    mass, 59
    Phoebe, 155
    radial spokes, 63
    revolution, 59
    rings, 59, 62, 63
    rotation, 59
    satellites, 60, 63, 64
    space probes, 60
    Voyager, 61
Sciaparelli, Giovanni, 32
Secci, Angelo, 32
Solar Flare, 144, 145
Solar System, 2
    origin, 151-155
    scale, 4
    structure, 4
    supernova genesis theory, 152, 153
Solar wind, 86, 137, 144, 145
Spectroscopy, 2, 135, 136
Spicules, 144

Spin-orbit coupling, 18
Sporadic meteors, 90
Stars, 3
    stellar system, 5
Sun, 3, 13, 133-149
    chromosphere, 137, 144, 147
    composition, 138
    corona, 137, 144-147
    coronal holes, 145
    differential rotation, 143
    envelope, 137
    filaments, 148
    fusion core, 137
    future, 149
    hydrogen fusion, 148
    interior, 148
    photosphere, 133, 137-143
    photospheric granulation, 138, 139
    plages, 138
    prominences, 145, 148, 149
    radiation pressure, 88
    solar flare, 144, 145
    solar wind, 133, 137, 144, 145
    spectrum, 136, 137
    spicules, 144
    sunspots, 138-143
Sunspots, 13
    Babcock theory, 143
    butterfly diagram, 141, 142
    magnetic polarity, 140
    maunder minimum, 142
    origin, 143
    penumbra, 138, 140
    sunspot cycle, 141
    umbra, 138, 140
Surveyor, 114, 116
Synchronous Rotation, 18, 102

T

Terrestrial planets, 47
Tethys, 65
Tides, 129-131
    neap, 130
    spring, 130
    tidal evolution, 130, 131
Titan, 63, 65, 66
Titania, 70
Tombaugh, Clyde, 76
Triton, 74, 75, 155
Trojan asteroids, 84

U

Ultraviolet Waves, 134
Umbriel, 70
Umbra, 122, 138, 140
Universe, 5
Uraniborg, 11

Uranus, 67
- clouds, 67
- composition, 67
- discovery, 67
- distance from Sun, 67
- magnetic field, 68
- period of revolution, 67
- period of rotation, 67
- rings, 69
- Roche's limit, 69
- rotation, 155
- satellites, 70
- shepherd satellites, 69
- temperature, 67
- Voyager, 68

V

Venus, 13, 24
- atmosphere, 25
- distance from Sun, 24
- elongation, 24
- greenhouse effect, 25
- phases, 13, 24
- rotation, 24-25, 155
- space probes, 26
- surface pressure, 25
- surface temperature, 25
- surface, 26-31
- volcanism 28-30

Vesta, 83
Viking, 35
Voyager, 48, 50, 61, 68, 73

W

Wave Model, of Electromagnetic radiation, 135
Wavelength, 135
Weight, 14

X

X-Rays, 134

Z

Zodiac, 7